IDRC-182e

Resource Allocation to Agricultural Research

Proceedings of a workshop held in Singapore, 8–10 June 1981

Editors: Douglas Daniels and Barry Nestel

Cosponsored by:
International Federation for Agricultural Research and Development
International Development Research Centre

Contents

Foreword

Interest in resource allocation to agricultural research has been growing since the pioneering work of Griliches and others increased our awareness of research as an economic activity requiring the allocation of scarce resources. Much of the literature, however, has focused on untried theoretical models that require an investment of manpower and time that is unavailable to many research managers and policymakers in developing countries. These managers and policymakers have had limited opportunity to contribute to discussions on this issue and little of their practical experience has, as yet, been published. This workshop, held in Singapore from 8 – 10 June 1981, was therefore heavily case-oriented and geared to giving representatives from national programs an opportunity to exchange views on their practical experiences. The overall objective was to review the existing state of the resource allocation process for agricultural research in developing countries and to consider ways of improving this process.

The meeting was cosponsored by the International Federation for Agricultural Research and Development (IFARD) and the International Development Research Centre (IDRC). IFARD, an informal association of national agricultural research directors from developing countries, has identified resource allocation issues as an area of particular interest. The IDRC has provided small grants to allow the preparation of inventories of research activities in a number of developing countries and was interested in reviewing the value of these and in receiving advice on what further support, if any, IDRC might provide. All of the participants were from developing countries. Observers included representatives from IDRC and the International Service for National Agricultural Research (ISNAR), the new international centre that is most directly concerned with the organization and management of national research systems.

The IDRC wishes to express its appreciation to the participants, who all prepared papers and contributed actively to the workshop, and to the Singapore office of IDRC, which provided excellent administrative and organizational support. Michael Graham, in particular, provided a great deal of assistance in editing the papers.

The specific objectives of the workshop were to review national inventory studies and assess their value; explore existing allocation systems in developing countries and suggest possible improvements; and identify further work that should be carried out to improve resource allocation. The workshop had five sessions: inventory presentations; allocation models; existing allocation systems; manpower planning; and a review of needed follow-up work. The discussions and conclusions of the meeting have been summarized and are presented at the beginning of the publication. The inventory papers, which follow immediately after the summary of discussions, present only a summary of the much more detailed reviews carried out in most of the countries represented.

The workshop focused on the allocation process at the sectoral and commodity level. It did not review multisectoral allocation issues or the need for better operational classifications and analysis of research expenditure and manpower allocation for use by the managers of research stations or project leaders, although this was an issue identified as requiring more analysis.

The participants agreed that sectoral resource allocation is becoming more complex as the number of research programs increases and as research systems become more diversified through such means as the creation of single commodity research institutes. Participants identified a number of issues requiring further study including a more systematic development and use of resource inventories. Some of the participants have begun to carry out further review and study of those issues outlined in the summary of discussions.

Although participants were drawn from nearly all developing country regions, it was felt that future collaboration might best be organized on a regional basis. There was a strong feeling, however, that follow-up activities should be undertaken by national research programs to ensure that the results would be specifically useful to national agencies. As IFARD represents the interests of national research directors, it was felt that some follow-up work might take place within the auspices of IFARD.

It is hoped that the papers and discussion summary included in this publication will prove useful and stimulating to researchers and policymakers in other developing countries. Agricultural research systems must assume a greater role in increasing agricultural production. At the same time, research resources are often limited and it is important not only to increase these resources but to maximize their impact through an effective allocation process.

Doug Daniels
Acting Director
Office of Planning and Evaluation
International Development Research Centre

Participants

Apinan Phatarathiyanon, Department of Technical and Economic Cooperation, 962 Krung Kasem Road, Bangkok, Thailand

Ekramul Ahsan, Member Director, Bangladesh Agricultural Research Council, Farm-Gate, Airport Road, GPO Box 3041, Dacca, Bangladesh

Fernado Chaparro, Regional Director, Centro Internacional de Investigaciones para el Desarrollo, Apartado Aéreo 53016, Bogota, D.E.

Doug Daniels, Acting Director, Planning, International Development Research Centre, P.O. Box 8500, Ottawa, Canada K1G 3H9

S.M. Elias, Head, Division of Agricultural Economics, Bangladesh Agricultural Research Institute, Joydebpur, Dacca, Bangladesh

Maria Aparecida S. da Fonseca, Centro de Fertilizantes, Instituto de Pesquisas Tecnológicas, Caixa Postal 7141, São Paulo, Brazil

Michael Graham, Regional Liaison Officer, Communications Division, International Development Research Centre, Asia Regional Office, P.O. Box 101, Tanglin Post Office, Singapore 9124

Fred Haworth, Senior Research Officer, International Service for Agricultural Research (Netherlands), P.O. Box 93375, 2509 AJ, The Hague, Netherlands

F.S. Idachaba, Department of Agricultural Economics, University of Ibadan, Ibadan, Nigeria

Nik Ishak bin Nik Mustapha, Research Officer, Economic Branch, Malaysian Agricultural Research and Development Institute, Bag Berkunci No. 202, Pejebat Pos Universiti Pertanian, Serdang, Selangor, Malaysia

Jingjai Hanchanlash, Regional Director, International Development Research Centre, Asia Regional Office, P.O. Box 101, Tanglin Post Office, Singapore 9124

Aida R. Librero, Director, Socio-Economics Research Division, Philippine Council for Agriculture and Resources Research (PCARR), Los Baños, Laguna 3732, Philippines

Stachys N. Muturi, Science Secretary, National Council for Science and Technology, P.O. Box 30623, Nairobi, Kenya

Chris MacCormac, Research Officer, Agriculture, Food and Nutrition Sciences Division, International Development Research Centre, P.O. Box 8500, Ottawa, Canada K1G 3H9

G. Montes, Departamento Nacional de Planeacion, Calle 26, No. 1319, Bogota, Colombia

Malik Mushtaq Ahmad, Senior Documentation Officer, Pakistan Agricultural Research Council, P.O. Box 1031, Islamabad, Pakistan

P.C. Munasinghe, Deputy Director, International Development Research Centre, Asia Regional Office, P.O. Box 101, Tanglin Post Office, Singapore 9124

B.L. Nestel, 38 Hatchlands Road, Redhill, Surrey, England

Luis Paz, Luis F. Villaran 383, Lima 27, Peru

Y.D.A. Senanayake, Dean of Agriculture, Faculty of Agriculture, University of Peradeniya, Peradeniya, Sri Lanka

Ramesh Sharma, Agricultural Projects Services Centre (APROSC), P.O. Box 1440, Lazimpat, Kathmandu, Nepal

Rungruang Isarangkura, Chief, Agricultural Planning Sector, Economic Projects Division, National Economic and Social Development Board (NESDB), 962 Krung Kasem Road, Bangkok, Thailand

Sjarifuddin Baharsjah, Director, Center for Agro-Economic Research (AARD), Juanda 20, Bogor, Indonesia

Eduardo Trigo, Coordinator, Cooperative Research Program on Agricultural Research in Latin America (PROTAAL), Apartado Postal 55, 2200 Coronado, San José, Costa Rica

Anthony Tillett, Associate Director, Social Sciences Division, International Development Research Centre, P.O. Box 8500, Ottawa, Canada K1G 3H9

Ramon Valmayor, Director-General, Philippine Council for Agriculture and Resources Research (PCARR), Los Baños, Laguna 3732, Philippines

F.J. Wang'ati, Secretary, Agricultural Sciences Advisory Research Committee (ASARC), Ministry of Agriculture, P.O. Box 30028, Nairobi, Kenya

Mohd. Yusof bin Hashim, Deputy Director-General (Research), Malaysian Agricultural Research and Development Institute, Bag Berkunci No. 202, Pejabat Pos Universiti Pertanian, Serdang, Selangor, Malaysia

Discussion and Conclusions

In many developing countries, agriculture is the dominant economic activity and principal source of both economic growth and export earnings. In recent years, much of the growth that has taken place in the agricultural sector has arisen from extending low-technology systems onto expanded areas under cultivation. However, the scope for further increasing the cultivated area is often limited. In many cases, the position has now been reached where it is necessary to raise production through increasing yield levels rather than by expanding the area of land under cultivation. Such a measure will necessitate the introduction and use of higher technology and more intensive production systems. Effective programs to do this require strong support from extension and research services. In the last decade, a number of governments have taken steps to provide this and there has been a marked strengthening of national agricultural research services.

Historically, such services have often been somewhat isolated from overall development goals and research has been carried out by a multiplicity of agencies and in a manner that has led to fragmentation and duplication. Furthermore, research has usually been organized on single crop or single discipline lines, although the subsistence farmer, who is the client for much of this research, does not often practice extensive monoculture but usually manages a complex system of intercropping. As a result of these circumstances, many research projects have had only a limited relationship to small-farmer needs.

Part of the reason for this is that most agricultural research activities were started as a sideline in the course of providing general agricultural services, particularly those given to support cash or export crops, and protecting imported exotic livestock from diseases and parasites. Indeed, even today only limited research has been carried out on some of the major subsistence food crops of the tropics.

Another reason for the disorganized state of agricultural research is the general lack of national research policies that specify priorities so that these can be used to effectively determine the most appropriate allocation of available resources. Thus, although many countries have had agricultural research services for periods approaching 100 years, little has yet been done in the way of establishing central inventories of research projects and programs. The value of the resources (manpower, land, equipment, and funds) devoted to individual programs and projects is seldom known. As a consequence, it has not been easy to provide appropriate information on the cost-effectiveness of agricultural research services. Although the literature contains a number of case studies indicating a high payoff to agricultural research, this information is not necessarily usable by the national planner who has to make decisions on the overall budget for agricultural research nor is it necessarily relevant to the task confronting the sector planner who is concerned with specific allocations within the agricultural research budget.

In recent years, a number of research managers and policymaking authorities have started to examine this issue with the objective of devising a more rational approach toward resource allocation for agricultural research. A prerequisite for this task is an adequate data base. In the first session of the workshop, a number of authors reported on their experiences in the preparation of country data-base inventories.

Inventory Studies

The inventories of agricultural research resources presented at the workshop represented either the first or the most comprehensive review of this kind ever carried out in these countries. The inventories focused on two resources, finance and manpower, classified in a number of different categories. Nearly all of the papers incorporated a comparison of research resources devoted to individual commodities or agricultural subsectors, with quantitative indicators of the importance of each commodity. This allows an initial assessment of the appropriateness of commodity resource use relative to the significance of each commodity. Table 1 summarizes some of the information on financial resources presented in 13 of the country studies (two African, two Latin American, and nine Asian countries). Some degree of standardization was attempted but proved difficult as the availability of information was so variable. In addition, the country studies in Kenya, Sri Lanka, and Bangladesh are not yet completed. Therefore, the information collected in the studies and summarized in the table shows a wide range in coverage and in the degree of disaggregation.

The collection and maintenance of an adequate inventory of all resources used in agricultural research was regarded as an essential first step in developing a more rational system of planning and resource allocation to agricultural research. An analysis of inventory data cannot by itself indicate how resources should be allocated but can illuminate problems in the research system that should be examined in more depth.

Although the central theme of the workshop was how to improve the allocation and increase the effectiveness of *existing* resources, it was also felt that national inventory data, supplemented by cross-country comparisons, and cost-benefit studies could highlight resource constraints and neglected research activities and thus provide a better basis for justifying *increasing* total resources for research.

Even though most studies required further refinement, and much of the information presented at this meeting did not lend itself to comparative analysis, the papers indicated several areas where problems common to a number of countries were identified and to some extent their magnitude quantified.

There was general agreement that most countries still lack an adequate system for planning, allocating, and monitoring research resources. As a result, there is excessive fragmentation and overlapping of research activities between different ministries and institutions and a misallocation of resources. Research activities often bear no apparent or consistent relationship to the economic or social importance of different commodities, the potential impact of such research on farmers and other clients, or national development and other political objectives.

Despite the increasing recognition given to the importance of agricultural research in achieving development objectives, some of the country papers indicated an apparent decline in real terms in research expenditure. Most countries, however, have increased the proportion of GDP and especially agricultural GDP devoted to agricultural research in the last decade. The relative level of resources devoted to research in the small number of countries represented at the meeting does not appear to bear any relationship to per capita income levels or the importance of agriculture in

Table 1. Agricultural research expenditure in different countries.

Country	Year	GDP	Agric. GDP	Value added by subsector		Value added by selected important commodities[a]	
				Public sector agricultural research as a % of			
Bangladesh	1979/80	0.21					
Brazil	1978	0.11[b]	1.19				
Nigeria	1975/80		0.004[c]				
Pakistan	1980/81	0.08[d]	0.25[d]				
Sri Lanka	1980	0.15	0.64				
Nepal	1979/80	0.17	0.26	Crops	0.40		
				Forestry	0.24		
				Fisheries	0.17		
				Livestock	0.02		
Philippines	1979		0.47[e]	Fisheries	0.87		
				Forestry	0.47		
				Livestock	0.37		
				Crops	0.37		
Thailand	1979	0.07[b]	0.27	Forestry	0.31		
				Crops	0.31		
				Livestock	0.14		
				Fisheries	0.10		
Colombia	1976		0.16			Wheat	1.42
						Cocoa	0.62
						Rice	0.08
						Cotton	0.05
Kenya	1979	0.39	1.14			Livestock	1.00
						Coffee	0.68
						Tea	0.20
						Wheat	0.003
Malaysia	1979	0.20[b]	0.79			Beef	8.94
						Rice	2.32
						Rubber	0.60
						Palm oil	0.12

Source: Data were derived by the editors from workshop papers in which tables were presented in different formats that are not strictly comparable. The figures need to be interpreted with considerable care.

[a] Figures shown are meant to show the range among commodities.

[b] Agricultural research expenditure expressed as a percentage of GNP.

[c] Planned agricultural research expenditure (at the federal level only) divided by value added by agriculture.

[d] Calculated by assuming a 7% increase over 1977/78 agricultural research expenditure.

[e] Calculated from 1977 GDP and 1979 research expenditure.

the economy. Discussions on the classification of research activities by subsector, as shown in Table 1, indicated that crop research was almost always given relatively more resources than its economic importance warranted and fisheries and forestry usually less, with animal research varying considerably. There was also, usually, a relatively high allocation of research resources for cash and export crops (especially where research resources were obtained through a producers cess or export levy). The reasons for such apparent anomalies are often historic.

11

Scientific manpower employed in agricultural research has shown a substantial increase in most of the countries represented, although the paper by Ardila, Trigo, and Piñeiro documents how dramatically research programs can disintegrate without continuing allocation of resources for training and adequate inducements to keep scientists. The studies provide evidence of these increases as well as some of the extreme variation in the kind of manpower available by level of training and discipline. This high degree of variation was evident between commodity programs within the same country and for the same commodity programs between countries. With the increase in manpower resources generally outstripping the rate of growth of budgets, the problem of ensuring adequate funds for scientists' operating requirements has been exacerbated. Papers presented on Bangladesh and Pakistan classify research expenditures by function and show that the availability of nonrecurring operating funds has dropped to less than 10% in many cases.

Inventory Methodology

Participants felt that no single classification system would suit every country and some degree of flexibility would be desirable at this stage in developing the methodology. Furthermore, different users had different data requirements. Some methods of classification would be useful for ensuring support from policymakers for agricultural research, whereas other and more disaggregated data are necessary for research managers. There are, as yet, no generally accepted standard *indicators* for classifying research activity. Nevertheless, a comparison of the various inventory studies under way and completed provided useful methodological information for refining what was regarded as a potentially valuable tool for both research managers and policymakers.

All of the country studies classified research activities on a commodity basis. There was considerable support for such a classification on the grounds that it was easy to prepare and of immediate use. It was recognized, however, that a commodity classification may not be useful in readily identifying research activities directed to planning and development objectives that have a strong socioeconomic element, such as farming systems, integrated rural development, and transmigration programs. Likewise, it was not useful in identifying basic, applied, and operational research activities; nor did it necessarily relate to the institutional distribution of research funding, which was the basis upon which both expenditure and staff were often budgeted. Nevertheless, at this stage of development of the inventory methodology the commodity approach appeared to be worth pursuing because it was used so widely. It was a particularly useful tool for cross-country comparisons, although these needed to be interpreted with a great deal of care.

Various ways in which research resources could be categorized are illustrated in Figs. 1 and 2. As pointed out in the captions, some of these categories probably have limited practical value, whereas other categorizations are almost impossible to calculate. Similarly, only some of the possible cross classifications illustrated would be useful, such as classifying research expenditures by commodity and source of funds when examining the direction of external agency research funds. Because there are many different ways in which research activities might be classified, it was agreed that data should be collected at the most disaggregated level possible in order to allow it to be reclassified in more than one of the different categories shown in Figs. 1 and 2. Given the diversity of possible classifications, participants stressed the importance of first defining precisely what should be inventoried and how the data collected would be used. Before an inventory was prepared it appeared desirable to ensure that the

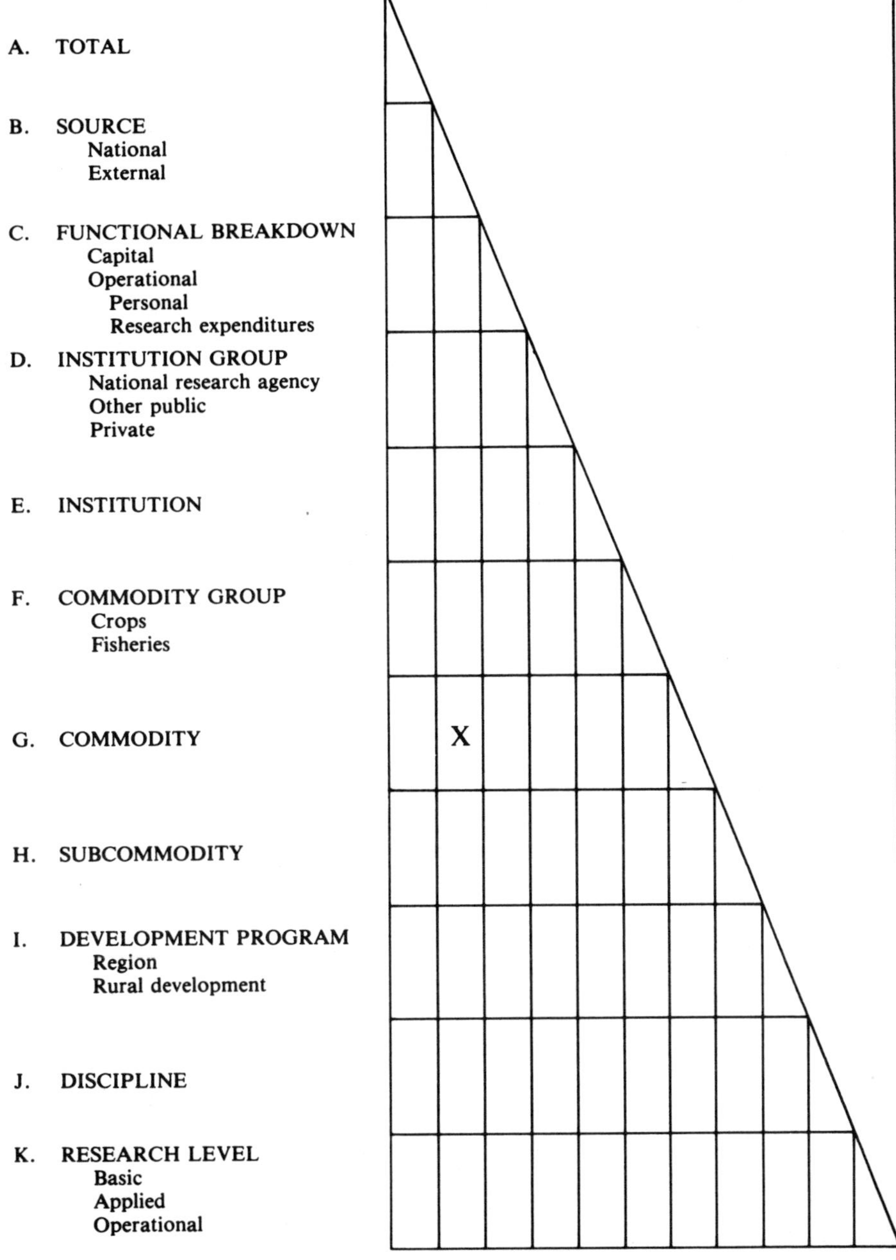

Fig. 1. Agricultural research inventory classification: expenditure. The matrix is meant to be illustrative. No one country determines expenditures in all the categories A to K and not every category is necessarily useful. In general, however, A to C are considered very useful; D to H are useful and possible to collect; and I to K are usually extremely difficult to calculate or have debatable value. The boxes within the matrix to the left of the diagonal line represent logically possible cross calculations. Thus, calculating G by B, in the square marked X, would provide a table showing both national and external financing for each commodity being examined. Calculating H by A would provide another table showing total research expenditure for subcommodity programs such as dry-land, irrigated, and deep-water rice.

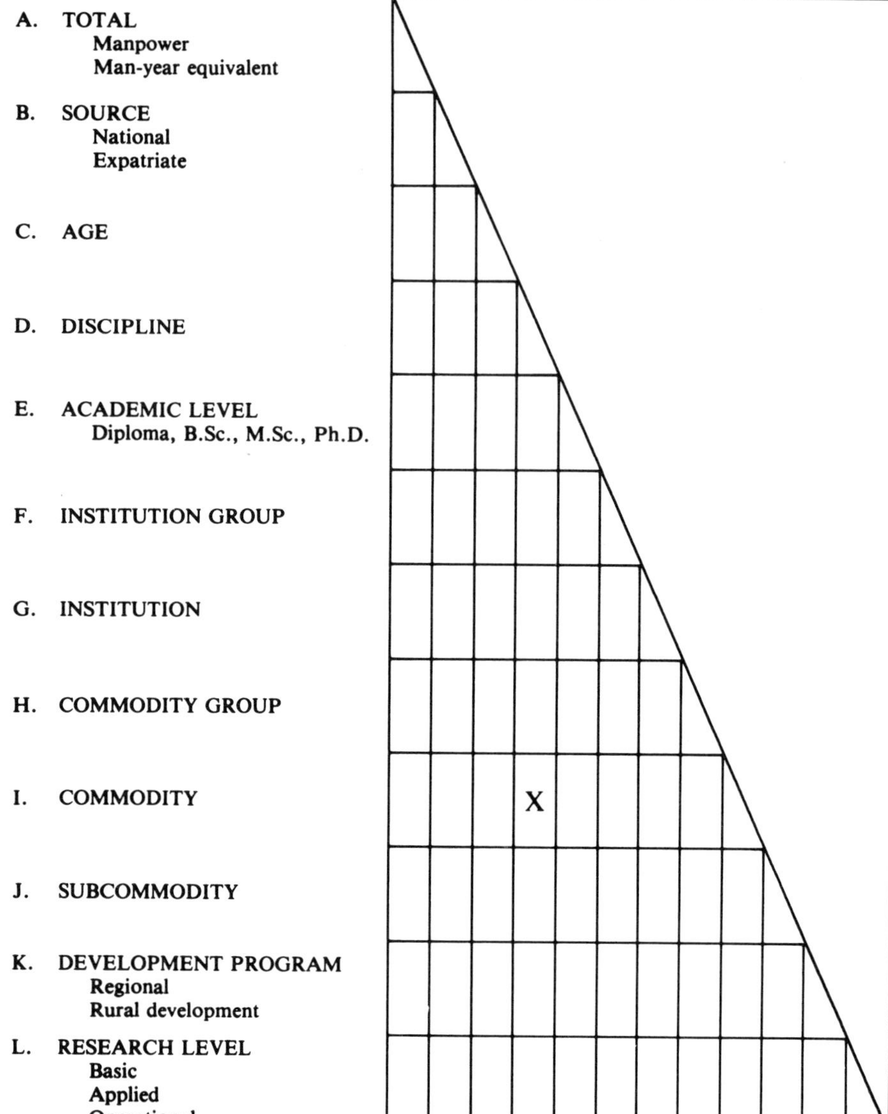

Fig. 2. Agricultural research inventory: manpower. The matrix is meant to be illustrative. No one country is classifying manpower in all the categories A to L, nor is every category necessarily useful. The boxes within the matrix to the left of the diagonal line represent logically possible cross calculations. Thus, calculating I by D, in the square marked X, would provide a table showing the availability of scientists by discipline for each commodity being examined. It was suggested at the workshop that a matrix showing estimated manpower requirements by discipline for each commodity program would be more useful for developing better estimates of future manpower requirements than simply asking research directors to provide estimates of total future manpower requirements for their institute.

data obtained would be: (1) easy to classify, (2) useful for cost-effectiveness studies, (3) operationally relevant in terms of the existing research system, and (4) easy to use for budgeting purposes.

The methodology used for data collection varied between countries. The Thai study conducted a comprehensive examination of government plans and programs and compared the research and manpower budgets in these with those available from key research institutions. By an iterative procedure a set of national tables was prepared. In Nepal and Malaysia, information was obtained from research agency documents and from questioning scientists. In Pakistan and Kenya, the bulk of the information was derived by submitting detailed questionnaires to research directors and their staff.

It was agreed that the use of questionnaires completed by research workers has some merit, although this approach suffers from the limitation that many scientists have problems in making subjective judgments regarding the portion of their time devoted to research. Published plans and programs also suffer from the limitation that they are usually prepared in advance and actual resource use often differs substantially from that planned. Station directors and administrators may be the best source for determining cost data, particularly when the results of questionnaires are compared with audited accounts from the previous year.

Another methodological issue discussed was the time period over which inventory data should be collected. The base year is likely to vary a great deal from country to country, with the most suitable base year for trend studies being the date of independence in one case or the establishment of an autonomous national research institution in another. Given the common use of 5-year plans, it was suggested that data be analyzed over these 5-year periods. Although it is useful to collect information on resource use patterns over a long period of time, particularly for use in studies on return to investment, it was agreed that it is extremely difficult to collect such information retroactively. Hence, it is important to collect such information as frequently as possible.

A number of countries participating in the workshop are now maintaining inventories on an ongoing basis. This is preferable to carrying out periodic surveys, although ongoing collection necessitates a ''centre'' that has the power to enforce reporting and the capability to process the data. It also requires computer facilities, which fortunately are now available in most national research centres. The volume of data being acquired globally suggests that there might be some merit in an international service serving as a world collecting centre, although for this to be of any value it would probably require a further degree of standardization of the inventory information.

Determining the True Costs of Research

The determination of total national expenditure for research presents problems in many countries. For example, time-series studies are often complicated by a high rate of inflation. To compensate for this it was suggested that research expenditure should be expressed in *constant* as well as in *current* cost terms.

It is often difficult to obtain research cost data from the private sector and from peripheral public sector organizations. Even in the public sector the task is not always easy. There are usually a large number of government ministries involved in agricultural research. In many cases, public agencies undertaking research also fulfill extension, development, or regulatory functions and it is not always possible to disaggregate expenditures between these various functions. Even in an agency concerned only with research, a substantial part of the budget is devoted to administration. Although it was recognized that this is an essential service component to support research, it is important to be able to identify it as a cost centre in order to establish the precise structure of the research outlay.

There are sometimes large discrepancies between the budget allocations for research and the actual outlay. Examples were given where the funds received were less than half those shown in the budget and for this reason it appeared important to express costs in *actual* rather than *budgeted* terms. The Colombian study indicated that over a period of years actual expenditures ranged from 70–90% of the budget. The study used 80% of the historical budget data to correct for this in its time series. National accounts in Colombia are now being adjusted to use actual rather than budgetary figures. Serious problems can arise unless national accounts and inventory studies based on detailed assessment of on-station expenditures can be made compatible.

It would be highly desirable to separate capital and operational expenditures for research. Capital grants from external donors often mask the low level of operating funds available for scientists and give an erroneous impression that the research budget per man-year is adequate. In fact, the lack of operational funds has been a major constraint. In some countries, manpower and essential services accounted for over 90% of the research budget, leaving little for actual research expenses. This was particularly true in universities, where research funds were often negligible.

Apart from examining capital *flow* on an annual basis, there is considerable merit in assessing the total capital *stock* in order to know whether there is adequate physical plant and equipment with which to conduct the planned research. In Kenya, such an inventory is being established in order to identify the location of specialized and costly equipment that might be used to service more than one institution. Knowledge of the total investment in specialized equipment is also useful for preparing a national depreciation schedule, which can be valuable for justifying budgetary requests for replacements, because many institutes do not depreciate their physical assets in their annual accounts.

Participants also discussed the difficulties encountered in trying to calculate indicators of the importance of different commodities. Production values, particularly for crops consumed on the farm, are often grossly underestimated in national accounts and thus lead to a downplaying of their importance when setting commodity priorities. Commodities often have different grades with different prices. Where commodity prices are artificially distorted, it was suggested that shadow prices should be calculated and used. Most of the inventory studies compared research expenditure and manpower per commodity with value of domestic production. It was suggested that commodity imports also be included to give a better measure of the real importance of each commodity to the country.

Classifying Scientific Manpower

The discussion on classification in inventory studies was continued in the fourth session dealing with manpower planning. Manpower data are often readily available but difficult to interpret. Many scientists, such as senior research managers, are not undertaking research or spend little of their time on research. Therefore, it is important to try to identify actual scientist man-years devoted to research (man-year equivalent) as well as the total number of scientists employed in research institutions. Participants noted the value of calculating level, discipline, and age of scientists. The inventory papers documented an enormous range in the relative levels of scientists in terms of Ph.D., M.Sc., and B.Sc. ratios between countries and commodities within the same country. Some skepticism was expressed about whether appropriate ratios could ever be developed but it was felt that this information can be useful in identifying where apparent imbalances exist. The Indonesian paper demonstrates the value of examining the age profile of scientists in research in terms of future training requirements.

There was some discussion on the value of going beyond quantitative values to incorporating measures of quality. The Kenya inventory study, which is still in progress, is using an index that weighs manpower according to level of training and working experience. A more comprehensive approach would be possible by using individual assessments such as the evaluation committee approach in Indonesia, which rates all scientists in terms of research, publications, and other scientific activities in an annual promotion system based on a point score.

Manpower for Research

Three papers on manpower issues were presented in this session. It was suggested that manpower should be treated as a primary input to the research system rather than as a residual as is the case in some countries. It was noted that major training programs are under way in Brazil, Malaysia, the Philippines, Indonesia, and Bangladesh. Training programs to meet estimated future manpower requirements were reviewed in papers on Indonesia and Bangladesh. However, even with large-scale training programs, the unattractive salaries and working conditions of agricultural researchers in many countries represent a major constraint and often lead to a serious brain drain in trained manpower. This was illustrated by an in-depth study by Ardila, Trigo, and Piñeiro on Argentina, Colombia, and Peru.

A wide-ranging discussion on manpower, which followed the three lead papers, dealt with five main areas: manpower requirements, manpower supply, links between the research system and the university, manpower management, and links between manpower planning and the national agricultural research plan.

It was felt that considerable work is necessary to devise better methods for determining *manpower requirements*. Future manpower requirements are usually estimated by asking institution directors for their subjective estimates and then aggregating these to develop a national target. In theory, it should be possible and would be preferable to estimate manpower requirements by developing a matrix (manpower input to research output) or building economic demand models. In practice, this type of approach has been looked at in some Asian countries but it appeared to make limited sense in terms of the funds available for training and supporting researchers. In both Nepal and Bangladesh this type of planning exercise came up with numbers that appeared unrealistically high in terms of the budget likely to be available.

In addition, it is questionable whether it is meaningful to determine the requirements for highly trained manpower unless this can also be related to the availability of nongraduate support staff for the researchers and extension staff, who would ensure that research findings were made available to the farmer. In some countries an imbalance is appearing, with the shortage of support staff constraining the value of output from highly trained research workers.

The discussion on manpower *supply* focused on funding restrictions and availability of new graduates, the long gestation periods involved before supply can be increased, the need to relate university programs to actual requirements, and the difficulties of reallocating existing manpower to newly defined priority areas.

Some papers calculated the level of output from national universities and the Indonesian paper reviewed how manpower requirements related to probable university output. In some cases, the universities are clearly unable to provide the levels of output required. In Africa, the universities are, in the main, relatively new and have limited facilities for research. For example, in Kenya only 1.7% of the national agricultural research budget went to universities in 1970. This makes it difficult for them to retain skilled manpower. A group of African scientists recently proposed to the Consultative Group for International Agricultural Research that donor support be given on a long-term basis to selected African universities to increase their training and research capital in the agricultural sciences. Without this long-term support it will be difficult to upgrade national agricultural research capacity.

The problems of relating university output with the needs of national research programs are accentuated in some countries where they are isolated from each other or even in competition. Certain countries have made great efforts to integrate the agricultural research and university systems. Sri Lanka was cited as one case where the university and Ministry of Agriculture are jointly represented on the Governing Council of a postgraduate Institute of Agriculture that determines the number and kind of students that will be enrolled. The costs of this training are borne by the Ministry of Agriculture. It also subcontracts research to academics to optimize the use of skilled resources. A closer working relationship is also being fostered in the Philippines where a number of Philippine Council of Agriculture and Resource Research (PCARR) Centres are located on campuses, with work being carried out jointly by teams comprising PCARR specialists and contracted faculty staff.

The discussion on management issues focused on how manpower is allocated within the research system, the scientist's role as a manager, upgrading of scientists, and the creation of a research environment in which the turnover of scientists can be reduced. One issue that was the cause of some concern was that of structuring appropriate incentives to get scientists to accept posting to remote duty stations with limited scientific, social, or educational facilities. Both PCARR and the Agency for Agricultural Research and Development (AARD), in Indonesia, have established selection criteria or special training awards to ensure that more students are trained in neglected research subjects or are willing to work in isolated locations. Closely related to this issue is the problem of how to persuade scientists to retrain or to shift their field of endeavour when their own field of interest and activity is downgraded in priority. Participants felt the issue of in-service training is a topic that has been largely neglected in many agricultural research services.

The high turnover of skilled manpower was another subject of widespread concern. Systems of bonding skilled personnel to recuperate the costs of their training are difficult to apply. Salary differentials with developed countries are so great that overseas posts represent an enormous temptation to ambitious young Third World professionals. International agencies are guilty of recruiting large numbers of skilled people urgently needed in their home countries. A paper from Latin America attempted to quantify the turnover in skilled agricultural scientists in key national institutions in Argentina, Colombia, and Peru. It showed that there was a critical mass of skilled scientists at the conclusion of major externally-funded schemes in all three countries but this steadily declined once external support for training ceased. Indonesia is trying to overcome the risk of such a situation by having most of its higher level training carried out locally. This involves a major strengthening of the graduate schools in its faculties of agriculture.

Undoubtedly, one of the key causes of the brain drain is funding instability, which not only keeps salaries low but results in erratic cash flows and budget cutbacks, both of which make the maintenance of a research program very difficult. However, there is not a great deal that the research scientists or institutions can do about these issues, which are essentially political in nature.

Many scientists who leave active research move into administration. This was felt to be beneficial in that the scientists concerned understand what research entails, but it was pointed out that these scientists usually have no training in research administration. In general, workshop participants felt that training in research administration is a much neglected activity and that most research managers would benefit from special administrative training. There appears to be a need for more flexibility in most research institutions to permit scientists to move back from management to research in the same way as deans revert from administration to teaching in universities. This involves questions of prestige and status as well as salary. It was suggested, but not generally accepted, that individuals be trained especially for management positions rather than promoting scientists who lack these skills. The answer probably lies in making management training an integral component of career development and in assessing managerial as well as scientific skills in the promotion system.

In summary, manpower planning in some countries is now an integral component of the national plan or strategy for agricultural research. Initial manpower planning has, of necessity, to focus on building up a critical mass of graduates at a certain level. Subsequent efforts to train disciplinary specialists have to take account of whether trained manpower can be attracted to crops, institutions, or localities that are defined as priority areas. The experience of countries where there have been large manpower training programs in the past suggests that a high wastage of skilled personnel is likely. This is particularly the case for disciplines such as economics and animal nutrition, where there is considerable scope for employment in the private sector. Manpower plans need to take into account the risk of staff turnover in such areas when quantifying training targets.

Defining Priorities

Four papers in this session reviewed the criteria and decision-making process for determining research priorities. The paper by Idachaba reviewed 12 criteria in some detail and there was general agreement at the meeting that these criteria were both relevant and important in determining research priorities and allocating resources. At the same time, participants agreed that conflicts will be inevitable if a large number of macroobjectives are established for agricultural research. Participants noted the difficulties in establishing priorities that take account of both equity and growth objectives, large and small farmer requirements, and short- and long-term research requirements. Responding to government priorities can create difficulties for the research system. Government policies can change rapidly and may unduly emphasize immediate needs to the detriment of a long-term research strategy. For example, it was suggested that the priority given to high-yielding cereal varieties has led to a relative neglect of research on other crops such as food legumes. This has had important nutritional implications in parts of Asia. Government equity goals that require emphasis on small farm development could reduce the economic impact of research programs and thus reduce public support for agricultural research.

The decision-making models presented focused primarily on using a combination of quantitative indicators of the importance of commodities and subjective assessments of the importance of different national research objectives. Only the Colombian paper outlined an alternate, more complex model. However, in none of these case studies (with the partial exception of the Philippines) was there enough experience with the procedures for defining either commodity or project priorities for their effectiveness to be adequately evaluated at the present time.

The paper from the Philippines described an ongoing allocation system. Each of more than 30 commodity research programs is designated at one of three priority levels. A set portion of the national agricultural research budget is allocated to each of these three levels. Commodity allocations at both the priority level and within this level are based on a series of socioeconomic, technical, and manpower criteria that were both objectively and subjectively assessed by the Philippine Council for Agriculture and Resource Research, on which technicians, academics, and cabinet ministers were all well represented. The priority allocation procedure is thus closely linked to the political system. Furthermore, the council, through disbursement of funds from its own budget and its right to approve all publicly funded agricultural research projects before they are submitted to the Budget Bureau, is thereby placed in a strong position to direct agricultural research. In most other countries, the agency responsible for agricultural research is constituted with rather less power and authority.

The three other case studies presented dealt with countries in which a change in the decision-making process for defining research priorities is still under consideration. In the case of Colombia, funding for agricultural research has fallen since the early 1970s and the mechanisms both for reversing this and for strengthening the process of defining research priorities were presented. The first entailed defining priorities in terms of food security or comparative advantage in an open economy. Although this model can incorporate changes in economic policy, it requires considerably more data as well as calculation and use of such concepts as ''shadow prices'' and the social costs of land, labour, and capital. The second model, based essentially on the market value of the commodity, both produced and traded, is simpler to calculate and uses readily available data. It is based on the assumption that total market value of each commodity is a sufficient proxy of the overall socioeconomic importance of each commodity that research priorities can be established based upon market values.

Another approach is currently under study in Peru where 13 priority criteria have been identified. Each of these was given a subjective weighting factor and combined with usually quantitative weights for each commodity to provide an overall ranking for 53 crop and 16 livestock products. Many of the weighting criteria used were common to those used in the Philippines.

The Colombian and Nigerian papers incorporated sections on determining an appropriate division in funding of research between the public and private sectors in the case of Colombia and federal–state funding of research in the Nigerian paper. One of the Colombian models uses price elasticities of demand to determine whether the public or private sector should finance research on a commodity. The Nigerian paper suggests research responsibility for basic and applied research by federal and state research organizations be determined on the basis of state financing of location-specific applied research and federal funding of basic research with its positive externalities or spillover benefits over the whole country.

During discussion, participants noted that input from the small farmer, who is often the primary client of the research system, in the decision-making process may be very limited, especially as farmer organizations are often weak. In such circumstances, research priorities should be specifically reviewed to ensure they are compatible with the small farmers' needs.

The potential payoff from further commodity research is an important factor. Tea in Sri Lanka and rice in the Philippines were cited as cases where the development of new technology, leading to substantial increases in production, made it possible to reduce research efforts on these commodities. However, in both cases a core of research activity has been retained rather than phasing out research on it completely. Thus, there remains a residual expertise on both crops should research on them need to be scaled up in the future.

One of the weaknesses of the allocation models discussed at the workshop was that they did not pay much attention to the priority that should be given to *opportunity* or *nontraditional* crops. Another factor that limits the value, especially of more complex models, is the weakness of available statistics, particularly related to crops consumed on the farm. Improving the reliability of such data was regarded as an essential step in efforts to develop more effective allocation of research resources.

The information presented in the papers indicates that few abrupt shifts in funding of different commodity programs have taken place. Participants discussed the need to ensure that shifts in priority allow sufficient time to permit a gradual change in the research activities of scientists and the development of appropriate delivery systems for new research activities.

Limited discussions took place regarding the definition of priorities within commodities. This may be a topic that is somewhat location-specific and difficult to discuss in generalized terms. A procedure for defining subcommodity priorities has been established and is being used in the Philippines and a proposal to follow a similar review is under examination in Colombia. It appears, however, that at present the main emphasis in most countries still lies on defining priorities *between* than *within* commodities.

Allocating Resources

The fact that a systematic process of defining priorities is still at its formative stage in most countries probably explains why the four papers on resource allocation presented at the third session addressed themselves more to the institutional mechanisms for allocating resources than to the allocation process itself. However, in most of the countries represented at the workshop the system for allocating resources for research has either undergone change quite recently or is currently under review.

It was agreed that much further work needs to be done in assessing the actual allocation processes used in different countries because describing existing formal systems is often misleading. Many coordinating agencies and the allocation mechanisms in place exist in name only and have marginal influence on the way resources are actually allocated.

The case studies presented indicate that agricultural research systems are becoming increasingly complex and, in nearly all countries represented, agricultural research is being carried out by a surprisingly large number of government ministries. There are at least 9 ministries involved in Bangladesh and 11 in Sri Lanka. Coordination of these independent research entities is still very limited in most countries, although there has been some apparent increase in the establishment or authority of centralized agencies.

The countries represented a spectrum of institutions coordinating research ranging from traditional ad hoc mechanisms to some form of research council whose objectives are to coordinate all research activity and advise the government on the level of funding required.

The majority of countries have some form of national research council although, as in the case of Malaysia, the council's authority is often only advisory. It does not monitor or review agricultural research activities or have any influence on allocations to agricultural research. Other national research councils have more authority, ranging from an effective advisory and coordinating role but with no direct allocating responsibility to ones that grant them direct approval authority over projects or enable them to provide financial support. Kenya was cited as one case where the National Research Council and its Agricultural Sector Research Committee have developed authority through the political and scientific expertise represented on the council as well as the council's authority to prepare, for government approval, national science and technology budgets that include agricultural research.

Also increasingly common in the last 20 years are national agricultural research councils such as those outlined in the Bangladesh, Philippines, and Pakistan papers. Although the authority of most councils to allocate resources is still limited, some councils appear to be increasing their authority. The Bangladesh Agricultural Research Council (BARC) was cited as one case where the Ministry of Agriculture (but not yet other ministries) has recently agreed that BARC should review and give its approval to ministry research programs before they receive government funding. The PCARR in the Philippines is perhaps the most influential council, with all research proposals first being submitted to PCARR for approval before government funding is provided. The PCARR also possesses considerable financial resources of its own that can be disbursed. There was some agreement that stronger and more established councils, through their advisory function, are best placed to develop macrocriteria for resource allocation and represent a genuine attempt to separate this activity from complete government bureaucratic control.

Another major development that can provide a better review and allocation of funds for competing research programs has been to bring different research institutions together under one umbrella organization, usually exclusively devoted to agricultural research. The Brazilian Agricultural Research Company (EMBRAPA), organized as a completely autonomous research organization, was one of the cases reviewed at the workshop. It is not always easy to achieve this autonomy because, for political reasons, there is often a great deal of pressure to retain research in the Ministry of Agriculture. This has advantages in maintaining the link between research and the ministry, whose extension agencies relate closely to the farmer. Furthermore, research activities can be maintained through the support of what is often an influential ministry. On the other hand, it suffers from the disadvantage that the research agency budget may have to compete for funds within the Ministry of Agriculture and may lose out to other agencies within the ministry.

Allocation Process

Most of the papers and the discussion implied that whatever the mechanisms for allocating resources, the process by which resources are actually allocated is still ill-defined and often relies more on historical, personal, or political influence than on any formal criteria. Even where specific criteria are defined, they are often not effectively utilized. There was little time available for discussion, however, on how a more systematic process could be implemented, although the absence of resource data in these papers was again cited as one critical constraint preventing a more rational approach.

Research project proposals and budget estimates are, usually, still prepared by individual research institutes or research departments within a larger organization. Budgets are rarely based on detailed project costings but usually extrapolate past estimates. When aggregated, these estimates form the initial total request for resources. In all four of the resource allocation case studies presented at this session, the initial budget request goes through at least one review process by a higher body. This generally results in revisions of the budget and the manpower request. Because these revisions are usually downward, it was implied by many of the participants that the original requests are inflated in order to compensate for expected reductions. When government budget authorities allocate funds for research without drawing on any scientific review process, budget allocations are often arbitrarily adjusted to meet general fiscal objectives and not specifically related to research requirements and opportunities. Thus, the system of budgeting for research frequently appeared to be rigid and inflexible, especially for monocrop research institutes.

The general discussion included reference to the pros and cons of central versus decentralized decision-making. Highly decentralized systems potentially allow for a greater degree of flexibility but they also permit duplication. Furthermore, in their request for funds they may lose out to other ministerial priorities. Conversely, a highly centralized system can be inflexible but may offer less chance of duplication and more bargaining power to obtain its share of the total resources available for research. In both Kenya and the Philippines there is a centralized system but it was also felt that there is a good mechanism for dialogue at the lower level.

It was argued by some participants that the degree of centralization was not the critical factor in ensuring the effectiveness of the research system but that the latter was largely dependent upon the extent to which the extension service was linked with research. This important topic lay outside the discussion theme of the workshop but clearly warrants further attention.

One of the advantages cited of a centralized system was its potential for ensuring that the valuable support provided by external donor agencies and the international agricultural research centres (IARCs) is geared to both national research priorities and to the availability of resources. A number of participants felt that external agencies often base their support for agricultural research on the special expertise or institutions that they possess or on their personal or institutional contacts in the developing world. These do not necessarily relate to national priorities nor do these agencies have a strong built-in mechanism for becoming aware of these priorities. As a result of this, the use of national personnel as counterparts in externally funded programs does not always optimize the use of manpower. One suggestion made was that external agencies make a greater effort to at least notify any national research coordinating agency of projects they are developing with individual institutions. It was also suggested that external donor support might sometimes be more effective if it were to divert from its historic bias of channeling funding for capital development to providing greater support for operational activities in view of the often limited funds available for operating requirements.

The workshop participants felt that a stronger and more explicitly defined system of national priorities would help to obviate these problems. It was suggested that national agencies need to be better briefed on the way in which the IARCs and other external agencies allocate their resources. In this way, both national and international programs could become more complementary rather than competitive.

Universities are one group of institutions that are generally not within the effective purview of national research councils and that participants felt could play a more active role. It was suggested that universities could usefully perform basic or regional research functions that would complement other public sector research. At present, however, universities often lack the resources to participate more actively.

Because of the difficulty of providing adequate finance for agricultural research, participants also discussed the role of the private sector. Private sector agricultural research generally falls into two categories. The first is the provision of superior germ plasm, in which to some extent it is in competition with the international agricultural research centres. The second is crop research, funded by some form of export cess. In the latter circumstance, the research may be conducted by a government institution, possibly under the control of a commodity organization such as the coffee producers in Colombia. Because the cost of research tends to be only part of the export cess, it is usually possible to stabilize the flow of funds for research even when the commodity price fluctuates. Generally speaking, when growers are paying for research on a commodity, it can be expected that they will take some interest in and benefit more from the results. However, information is limited on the cost-effectiveness of this approach. Certainly, its funding appears to be more generous and stable than much public sector funded research. This continuity of funding as well as the possibility of tapping new resources for research justifies a much closer examination of the potential and value of private sector funding despite the lack of flexibility it introduces in resource allocation.

In general, participants felt that the issues raised in this session required much more investigation on the actual process used for allocating resources; how external funds and different national institutions such as single commodity, university, or private sector research could be better directed to national requirements; and how research institutions could influence the level and direction of resource allocation.

Next Stage

Participants felt the workshop was useful in providing information on resource allocation in other countries and exchanging ideas on different approaches to this issue. However, as these summary notes indicate, there are still considerable gaps in the kind of information available and a number of issues that need to be examined in more depth.

It was stressed that further work in this area, especially in the definition and development of resource inventories, should be carried out by national, not external, agencies. National agencies will be the primary users of such information and they need to increase their own operational competence in this subject. In most cases, the number of individuals within any one country concerned with the broader issues of resource allocation is still too limited. A broader base of interested individuals has to be created to provide an environment in which improvements in the resource allocation process can be introduced. This involves creating a better dialogue between policymakers and scientists to allow research institutes and their staff to influence and improve allocations. In this regard, one participant indicated that his inventory paper will be published and circulated in the national language and used as a basis for bringing together policymakers and the independent research institutions in the country to try to create some review and coordinating group.

At the same time, participants felt that there was merit in exchanging information and carrying out collaborative analysis by creating a network of interested researchers and policymakers primarily on a regional basis. Various regional organizations, as well as IFARD, were mentioned as possible coordinators of regional groupings. External donor agency support is probably necessary to allow these regional exchanges to take place.

Four priorities were stressed for follow-up activities:

(1) There is a need to improve the frequency, reliability, and scope of inventory studies in order to establish a battery of key indicators for research and development, particularly those that will be of value for cost-effectiveness studies and cross-country analyses. In order to do this, it is necessary to evolve a methodology that is simple enough to be easily adapted in all countries. Most participating countries encountered difficulty in defining and classifying research activities and there was considerable variation in what was included in different classifications. It was suggested that a working group be created to establish a standard definition and classification system for categorizing research activities.

(2) Regional networks should be created to allow exchange of information and experiences on common problems and interests. Some examples mentioned were techniques for evaluating scientists, the use of different criteria for establishing priorities, improvements to the resource allocation process, and alternate funding

mechanisms being developed to finance research. Such regional groupings would also allow for collection and analysis of regional resource-use patterns in agricultural research.

(3) Participants stressed the need for training seminars for national scientists on such issues as the management and administration of research or on the means by which research results can be disseminated to policymakers.

(4) Certain issues require in-depth case studies that could be identified and organized by these regional groups. Manpower planning, including better procedures for determining supply and demand and wastage rates, was cited as one example. Another topic that the meeting felt warrants early attention is the relationship between the research programs of national and international agricultural research centres. Several participants suggested that national research systems tend to be passive participants in any dialogue with the international centres. They felt there is a need for national programs to have a stronger influence on program formulation at the international centres. Although international centres might welcome such a dialogue, it would first be necessary for national programs to have a much clearer idea of their priority needs if this dialogue is to be effective. Another topic deserving more in-depth review is program evaluation, because few countries have clear ideas on how to go about this. A possible first stage would be to conduct in-depth analyses of past research programs that have been successful in order to try to identify what were the key elements for success. There is some literature available although little has yet come from developing countries. In addition, success has usually been evaluated in technological terms rather than in terms of the socioeconomic goals of most development plans.

Allocation of Resources to Agricultural Research: An Inventory of the Current Situation in Kenya

F. J. Wang'ati[1]

Although agricultural research has been conducted in Kenya for about 80 years, no comprehensive effort has been made to establish a central inventory of all research projects and programs and of the value of resources — manpower, land, equipment, and finances — devoted to research activities in various fields. The main reason for this is that almost all agricultural research activities were started as a sideline in the course of general agricultural services, especially in support of cash crops, such as coffee, tea, and sisal, and the protection of valuable dairy cattle from diseases and parasites. This approach is further exemplified by the almost total absence, until the early 1960s, of research on traditional food crops, like maize, sorghum and millets, beans, and potatoes, and the fact that up to the present time none of the government ministries has a scheme of service designed specifically for research staff. In the Ministry of Agriculture, all research scientists are still designated Agricultural Officers without distinction from the extension and regulatory staff. Problems have arisen with recruitment of nonagricultural graduates who are regarded as a nonprofessional cadre and accorded lower status. A similar situation exists in forestry where research has not even been accorded division status and in water development where formal research activities are only now being established.

Agricultural research has gained considerable impetus in the 17 years since Independence and the problems of coordination, logistic support, and control are increasing with every new research facility or program started. The current exercise of registering all research projects, programs, and their resources comes therefore at a vital stage of the country's development when severe competition among various national services for limited resources requires a continuous realistic assessment of the country's priorities.

Design and Administration of Questionnaires

Starting with no local experience whatsoever and guided mainly by a series of questions to be answered on a regular basis, it took us 5 months to design an appropriate series of questionnaires and a computer coding system to facilitate information storage, retrieval, and analysis. The administration of the questionnaire to all research institutions also proved a difficult and time-consuming exercise, especially since most of the research staff were not used to compiling such information. Research stations in Kenya also play a significant role in liaison, extension, and other services whose demands on the institutions' resources are both highly variable and difficult to quantify.

Most of the publicly funded research institutions have now been covered by the survey, but private companies and international organizations with research programs based in Kenya and the programs based at the university still must be documented.

Preliminary Results of the Research Inventory

Tables 1–4 show the preliminary results of the survey. The coding and computerization of the data must still be done and more detailed analysis of the information will increase and improve both the accuracy and consistency of the data.

Table 1 shows total agricultural research expenditure as a percentage of GDP for the year 1979–80. We plan to chart the trends in these figures for the period 1971–80 as the information is compiled. It is significant that although total research expenditure

[1] Secretary, Agricultural Sciences Advisory Research Committee (ARSARC), Ministry of Agriculture, P.O. Box 30028, Nairobi, Kenya.

Table 1. Total agricultural research expenditure[a] as a percentage of national research and development and as a percentage of GDP.

	(A) GDP[b] (K£, millions)	(B) National expenditure on research[c] (K£)	(C) Expenditure on agricultural research[c] (K£)	C/A (%)	C/B (%)
1970	512.51	396 607	391 507	0.08	99
1971	570.06	232 851	207 424	0.04	89
1972	666.22	1 422 138	1 405 711	0.21	99
1973	749.21	2 259 074	2 132 708	0.28	94
1974	907.63	3 031 945	2 901 101	0.32	96
1975	1057.22	3 287 108	2 931 955	0.28	89
1976	1278.10	4 259 433	3 668 383	0.29	86
1977	1640.65	8 279 410	5 726 292	0.35	69
1978	1788.41	8 936 422	6 374 553	0.36	71
1979	1974.97	9 509 032	7 010 672	0.35	74

[a]Crops, livestock, and range research.
[b]Gross domestic product at current prices based on economic surveys published by Central Bureau of Statistics.
[c]Based on government estimates of expenditure and therefore excluding expenditure by private companies and international institutions.

Table 2. Estimated research expenditure in relation to production values of agricultural commodities (1979–80).

	Estimated production value[a] K£('000)	Research expenditures[b] (K£)			Research expenditure as % of	
		Local	External aid	Total	Production value	Total research expenditure
Coffee	106 426	677 654	50 000	727 654	33.2	26.7
Tea	67 343	138 018	–	138 018	21.0	5.1
Maize	9 363	218 889	10 944	229 833	2.9	8.4
Wheat	14 886	39 205	–	39 205	4.6	1.4
Sugar	23 302	101 131	–	101 131	7.3	3.7
Other food crops	20 356	267 850	160 688	428 538	6.4	15.7
Oil and fibre crops	12 440	206 916	18 444	225 360	3.9	8.3
Horticulture	4 286	175 951	37 290	213 241	1.3	7.8
Livestock: beef and milk	61 890	405 191	94 350	499 541	19.3	22.8
Range research		88 170	34 945	123 115		
Total	320 292	2 318 975	406 661	2 725 636		

[a]Recorded marketed production in 1979, published by Central Bureau of Statistics.
[b]Based on survey of estimated expenditure on research projects.

in 1980 was less than 0.5% of GDP, at least 70% of this expenditure was devoted to agricultural research.

Table 2 analyzes the distribution of the research resources devoted to agriculture among various commodities and the relative values of these commodities. With the exception of coffee, tea, wheat, and sugar, which are centrally marketed, the values of the other agricultural commodities are very difficult to determine because only variable proportions are marketed through recordable channels. A good example is maize, which is the staple food for most Kenyans. It is estimated that approximately 1.6 million tonnes of maize were produced in 1980. At the official price of approximately K.Sh. 1/kg, the value of the crop should be in the region of K£ 80 million or nearly 10 times the published values of gross farm revenue. On the research expenditure side, with the exception of the principal cash crops like coffee, tea, and sugar, it has proved difficult to

Table 3. Scientific manpower engaged in agricultural research (1979–80).

	B.Sc.	M.Sc.	Ph.D.	Total no.	% of total
Coffee	6	11	4	21	6.9
Tea	1	2	1	4	1.3
Maize	5	5	0	10	3.3
Sugar	8	3	1	12	3.9
Wheat	9	2	1	12	3.9
Other food crops	55	24	10	89	29.1
Oil and fibre crops	14	4	0	18	5.9
Horticulture	16	8	0	24	7.8
Livestock	9	9	1		
Animal production and diseases	37	16	29	116	37.9
Range research	7	4	4		
Total	107	88	51	306	100

separate research expenditures for individual crops. Fortunately, a certain degree of specialization exists for groups of crops in various research stations and the survey data can be used to estimate proportions of funds devoted to such groups of crops and to a lesser extent the individual crop components. It is not uncommon to find one research officer in charge of a discipline covering several crops. In such cases, apportionment of time and experimental costs can be attempted but the results remain largely guesswork.

Notwithstanding the above difficulties, the survey data are already revealing important discrepancies between the proportional values of the commodities and the research resources devoted to them. Reasons for this situation are historical, but it is hoped that such analyses will help correct imbalances.

Table 3 shows the qualifications and distribution of research staff among various commodities. Generally, the quantity and quality of scientific staff seem to bear little relationship to the relative importance of the commodities. A good example is the small number of scientists working on the staple food, maize, in comparison with animal diseases. Most of the scientists with a Ph.D. and a good number of those with a M.Sc. are expatriates. These tables will be very useful in assessing the quality of research that can be expected in support of various commodities.

Table 4 gives the allocation of resources between research institutions. Here again, research resources are not allocated according to either the size of institution in terms of number of research staff or the number of stations served by each institution. It is suspected that the large and permanent overhead costs built up in stations will limit the capacity of the organization to respond adequately to various demands on its research services. In many cases, the institution budget has continued to grow year by year in spite of the reverse trend in quality of scientific staff.

Organization of Research

Agricultural research in Kenya is carried out under four government ministries, by international institutions, and by the private sector (Table 5). These bodies are free to decide what projects should be given priority as long as they can convince their financiers to allocate money. No central coordinating mechanism existed until the National Council for Science and Technology (NCST) and the Agricultural Sciences Advisory Research Committee (ASARC) were formed in 1977 and 1979, respectively. Allocation of resources therefore tended to depend on either urgent needs to solve crises in production or the skill of the research director in justifying expenditures to Treasury. The current project, creating a central register of all projects and resources allocated to them, is the first comprehensive exercise of its kind in Kenya. It is expected that the NCST and ASARC will use the data to formulate mechanisms to evaluate the cost-effectiveness of research projects and hence advise on the appropriate levels of resources that should be allocated.

Problems Encountered in the Survey Project

Design of Questionnaire

In the absence of previous local experience, the survey questionnaires were designed by first listing a number of questions we expected the survey to answer. The survey team was also keenly aware of the general reluctance of people, especially scientists, to fill in questionnaires and an effort was made to keep the questionnaires as simple and undemanding as possible. In this process, a problem arose in balancing brevity with clarity of the information requested and the need to administer the questionnaires personally to all scientists. Nevertheless, it was necessary to formulate brief separate instructions on how the questionnaires were to be completed. The design of the questionnaire was also influenced by the need to code and computerize basic information for mechanical analysis, retrieval, and updating. In this respect, qualitative information has been kept to a minimum but even this has created problems of appropriate coding.

Table 4. Allocation of government expenditure on research (1979–80).[a]

Institution	No. of research officers	Funds allocated		No. of stations
		£	% of total	
Crops				
SRD stations[c]	169	2 614 593	29.3	15
KARI — crops[d]	31	1 155 834	13.0	1
CRF[e]	21	41 000	0.5	3
Livestock				
VRL[f]	38	825 902	9.3	1
KARI — veterinary	35	847 792	9.5	1
Animal husbandry	40	485 172	5.4	3
Range research	16	512 850	5.7	3
Natural resources				
Wildlife	10*	520 964	5.8	3
Fisheries	3*	462 505	5.2	2
Forestry	6*	257 065	2.9	4
Health				
Medical research	13*	910 070	10.2	2
Trypanosomiasis				
Industry				
Industrial research and development	14*	290 506	3.3	1

[a]Based on government estimates of expenditures.
[b]Staff in post (1980). Numbers marked * signify total establishment.
[c]Scientific Research Division of Ministry of Agriculture.
[d]Kenya Agricultural Research Institute. Includes a small component of forestry research.
[e]Coffee Research Foundation, funded mainly by coffee industry. The figures shown represent only direct grants by the government.
[f]Veterinary Investigation and Research Laboratories. Responsible for both diagnostic services and research.

Response to the Questionnaire

Except in the university, where little progress has been made, the survey project has been well received in publicly funded research institutions. Scientists have seen a hope of future participation in decisions regarding the resources allocated to their projects and have therefore responded positively. This enthusiasm has however been frustrated by the fact that in most cases no records are kept on the expenditure in each project. Resource allocation is currently so centralized that most scientists have no idea how much money is available for their projects until resources run out and work is suspended. This not only affects individual scientists, but applies to some extent to large research stations that have on several occasions not been involved at all in either budgeting or defending their estimates to Treasury. The scientists, however, appreciate and welcome a simple procedure for realistic costing of their projects and even some training opportunities in this important aspect of their careers. It will take a long time to critically examine the costing of projects in the questionnaires, but any effort devoted to this process will be well rewarded in a simpler system for preparing estimates.

Follow-up Action on the Survey of Research Allocation

It is too early to predict the reaction of funding agencies to the current survey in Kenya, but the following procedures are likely to gain acceptance.

Central Register of Research Projects

The current survey will result in a central register of research projects and programs classified by commodity and institution. A centralized project coding system has been developed and will be adopted to facilitate monitoring and coordination. If a copy of this register is available in every research institution, it will encourage communication between scientists and avoid duplication of research effort. It will also be easier to identify projects that do not seem to be effective and reallocate resources accordingly.

Increased Awareness of Research Costs

The expected availability in Kenya, for the first time, of data on the actual distribution of resources to various commodities and the continuing effort to monitor value of commodities will bring into sharp

Table 5. Institutions conducting agricultural research in Kenya.

Ministry of Agriculture
    Scientific Research Division
    Kenya Agricultural Research Institute
    Coffee Research Foundation
    Tea Research Foundation
    National Irrigation Board
Ministry of Livestock Development
    Veterinary Research Laboratories
    Animal Husbandry Research Stations
    Range Research Stations
Ministry of Power and Communications
    Meteorological Service
Ministry of Higher Education
    Faculty of Agriculture, University of Nairobi
    Egerton College
    Kenyatta University College
International
    ICRAF (Base)
    ICIPE (Base)
    CIP (Outreach)
    ILCA (Outreach)
    CIMMYT (Outreach)
    ILRAD (Base)
Private
    Wellcome Laboratories
    Kenya Canners
    Other Companies

focus the need to increase cost-effectiveness of research. More rational budgeting and administrative procedures can then be adopted to correct imbalances in resource allocation and to make it easier to identify and quantify gaps that must be filled by technical assistance.

Further Surveys

Research is dynamic and project status changes depending on breakthroughs, change of personnel, or even change of priorities. It is therefore essential to constantly update the information collected in the current survey. The ASARC Secretariat in Kenya has been given this responsibility and it is expected that a new listing will be published each year. To maintain the necessary momentum and effect continual improvement on the information and method of presentation, it would be useful to establish a continuing forum of consultations between similar programs in developing countries.

I would like to acknowledge the input of my colleagues, S.N.Muturi, N. Mwara, W.M. Mwangi, and G. Ruigu, in the design and administration of the questionnaires. I also acknowledge with thanks the assistance provided by IDRC in launching the survey project and facilitating attendance at this workshop.

Inventory of Agricultural Research Expenditure and Manpower in Thailand

Rungruang Isarangkura[1]

In Thailand, about 35 million people, or 80% of the population, reside in rural areas and depend primarily on agriculture for their livelihood. The agricultural sector employs about 75% of the labour force and produces about 30% of the gross domestic product (GDP) and about 60% of total exports. The recent increase in agricultural employment (2% p.a.) was less than the increase in the rural labour force (3% p.a.) resulting in substantial rural–urban migration.

Over the past two decades, a major source of economic growth, and the most important element in the alleviation of poverty, has been the growth and diversification of agricultural production. This growth has been based on the extension of a low technology system of agriculture over the expanding cultivated area. A growth process based almost entirely on expansion of the cultivated area obviously cannot continue forever; in fact, the Kingdom may already have reached the end of its land frontier. It is critical for the government to focus on the opportunities for intensive agricultural development and on the policies and programs that will provide the agricultural sector with the necessary incentives and environment to realize these opportunities.

Agricultural Growth and Prospects

For most of the past two decades, agricultural GDP at constant prices has grown at about 5% per year. But, in recent years the overall growth rate of the agricultural sector has fallen to below 5% per year.

Crop production accounts for more than 70% of agricultural production and paddy is the main com-

modity. Livestock and fisheries contribute 10% each to the agricultural production value. Forestry, however, has lagged behind: there has been much illegal felling of timber and clearing of land for agriculture and little success in reforestation or the development of industrial plantations. The fisheries sector has done well because of the rapid growth of marine fisheries in the 1960s and 1970s. However, the 200 mile exclusive economic zones have caused severe reduction to the Kingdom's fisheries production in the recent past. Livestock production increased rapidly during the early 1970s primarily due to the expansion of pig and poultry production.

Expansion of the farm holding and planted areas has been accompanied by little change in agricultural technology. Farmers still use relatively small amounts of modern agricultural inputs and average crop yields have only increased an average of 0.5% annually over the last decade. The agricultural sector in Thailand is, therefore, faced with the need to increase output from the present farm area. There are significant possibilities, both for increasing the planted area and for raising crop yields. The provision of new irrigation facilities or the improvement of existing irrigation infrastructure would allow double or triple cropping. More important is the need to identify the constraints preventing more productive use of the rain-fed areas and to recommend the appropriate measures and rain-fed production packages to increase the planting of these areas. This would require developing technological packages relevant to disparate agroeconomic conditions, introducing appropriate incentives, including pricing incentives, and expanding supporting services, of which agricultural research is the key element.

Agricultural Research

Effective programs of agricultural research and extension are critical to the change from extensive to

---

[1] Chief, Agricultural Planning Sector, Economics Project Division, National Economic and Social Development Board (NESDB), 962 Krung Kasem Road, Bangkok, Thailand.

intensive agricultural development and, in recent years, the government has begun to give higher priority to the development of these services. Agricultural research is to be strengthened and decentralized to make it more responsive to local needs, and more emphasis is to be placed on developing optimum farming systems for rain-fed areas.

A major constraint to the development of an effective research system has been the lack of information on the allocation of resources to agricultural research under the existing organization and management of research services. It is generally believed that agricultural research has not been oriented to support the Kingdom's overall development policy, which attaches high priority to improving the income of the country's poor and thereby reducing income disparities. Agricultural research is carried out by several government agencies and, in some areas, the responsibilities for research are unclear, which results in excessive fragmentation and duplication of research activities. Furthermore, agricultural research has been organized along single crop or single discipline lines, has been heavily centralized in Bangkok, and has been unable to emphasize the development of farming systems appropriate to the agroeconomic conditions of the different areas of the country. However, because agricultural research plays a vital role in agricultural development, consideration must be given to overcoming these constraints. A prerequisite for sound policy analysis and planning is an adequate data base, and for a number of years planners have been concerned about the poor quality of the data base in Thailand.

Agricultural Research Institutions

Agricultural research is conducted mainly by public institutions because there are no private or international research organizations carrying out biological research programs in agriculture. Private commercial companies and bilateral and international organizations have provided financial support to the public research institutions. But, agricultural research is considered as a service provided by the public sector to the agricultural sector.

Agricultural research is carried out by several government agencies (Table 1). The most important ministry for crop, livestock, forestry, and fisheries production research is the Ministry of Agriculture and Agricultural Cooperatives, which has several departments supervising numerous up-country research and development stations. The Bureau of Universities encompasses Kasetsart University in the Central Region, Khon Kaen University in the

Table 1. Institutions conducting agricultural research in Thailand.

Ministry of Agriculture and Agricultural Cooperatives
    Department of Agriculture
        (85 experiment stations)
    Land Development Department
        (60 land development centres)
    Livestock Department
        (35 stations)
    Department of Fisheries
    Royal Forest Department
    Office of Agricultural Economics
    Office of Under-Secretary
        (4 regional offices)

Bureau of Universities
    Kasetsart University (Central Region)
    Khon Kaen University (Northeast)
    Chiang Mai University (North)
    Songkhla University (South)

Ministry of Education
    Agricultural Colleges (40)

Ministry of Industry
    Sugarcane and Sugar Institute

Ministry of Finance
    Tobacco Monopoly

Ministry of Interior
    Public Welfare Department
        (Tribal Welfare Research Center)

Ministry of Science, Technology, and Power
    Applied Scientific Research Corporation of Thailand

Northeast, Chiang Mai University in the North, and Songkhla University in the South, the leading academic research institutes in agricultural research. Other universities, such as Chulalongkorn, have recently become more involved in agricultural research in the context of rural development. The Applied Scientific Research Corporation of Thailand (now in the Ministry of Science, Technology, and Power) has concentrated on agroindustrial research. The Sugar and Sugarcane Institute in the Ministry of Industry conducts simple varietal tests and research on cultural practices for sugarcane; whereas, more advanced research on sugarcane is done by Kasetsart University and MOAC. The Ministry of Finance maintains eight plantation stations for tobacco. There are about 40 agricultural colleges in the Ministry of Education that conduct demonstration plots for college students, but their role in agricultural research at the present is not meaningful. Lastly, the Public Welfare Department in the Ministry of Interior has become more and more involved in agricultural development in highland areas in the northern part of the Kingdom. The department is spending close to B 2 million annually for agricultural work and has been increasing its staff in agriculture. In the past, it has not contributed to agricultural research but in the future it is possible that some highland agricultural research will be carried out. For the government budget, requests for funds for agricultural research are submitted to the National Economic and Social Development Board for review and passed to the Cabinet for approval. Annually, the Budget Bureau with approval from the Cabinet also provides the National Research Council (NRC) with a lump sum budget for research work to support programs determined by research committees at NRC. The total annual allocation is about B 10 million. Another source of funds is external grant assistance, which is principally managed by the Department of Technical and Economic Cooperation (DTEC) in the Prime Minister's Office.

Agricultural research projects have mainly originated from various departments located in Bangkok and because there were no clear ties between the researchers and extension workers there has often been no clear link between the project's objectives and local needs. In addition, there has been no effective mechanism for coordinating research programs. The points of coordination are the central agencies responsible for approval of the projects; however, national research policy has been rather general with no specific priorities to guide resource allocation. The country needs an agricultural research plan and must improve the research system to make research more area specific and relevant to local needs.

Inventory of Agricultural Research Expenditure and Manpower

Methodology

Lists of agricultural research projects were obtained from various government and nongovernmental agencies and the details of the projects were collected. Research institutes conducting agricultural research were identified and their projects were categorized according to research institute and commodity on which the research was performed. The initial analysis was followed by personal interviews of key persons in the institutes. Additional information on some projects was obtained by direct contact with the research directors of the projects. It was felt that the project approach in this study would avoid double counting of manpower engaged in agricultural research activities because agricultural researchers in the Kingdom are carrying out more than one project at any one time. In addition, senior agricultural scientists, including all project directors, have many responsibilities in nonresearch work.

Agriculture in this case covers agronomy, horticulture, fisheries, livestock, forestry, and related subjects and financial resources include both local and external sources. Research is classified as an innovative systematic activity undertaken to increase the stock of scientific and technical knowledge.

Because researchers are not engaged full time in any particular project, man-year equivalents are estimated for each group of scientists based on their positions in the projects, their qualifications, and their research institution.

Total Investment in Agricultural Research

During the fiscal years of 1974–79 the Kingdom has steadily increased its investment in agricultural research. Research expenditures have increased from about B 200 million in 1974 to B 400 million in 1979. The average research expenditure represents only 0.09% of GNP or 0.30% of agricultural GNP (Table 2). The level of investment in agricultural research is lower than the international standard: most developing countries are spending somewhat over 0.3% of their agricultural GDP on research and scientifically advanced countries spend about 1.0%. However, research expenditures in Thailand are higher than the average expenditures for South and Southeast Asia, which for 1975 were estimated at 0.2% of agricultural GDP. Even so, total investment in agricultural research is far below the target of 0.5% of agricultural GDP suggested by the World Food Conference.

The increase in agricultural research expenditures from 1974–1979 was at a higher rate than the increase in agricultural GNP over the same period. Similarly, the increase in research expenditures was higher than the farm population increase. Consequently, research expenditures per capita increased from B 6.2 in 1974 when the farm population was 32.8 million to B 11.2 in 1979 when farm population was 35.2 million. These figures are higher than averages that have been reported for Asia (B 3.1) and Asia and Southeast Asia (B 5.2).

The average total expenditure on agricultural research represented only 0.6% of the total public expenditure during 1974–79 with a maximum of 0.7% in 1975 and a minimum of 0.4% in 1979.

Source of Funds

The government is the main source of funds for agricultural research. External assistance constituted less than 0.4% of the total expenditures for

Table 2. Total agricultural research expenditure (millions of baht)[a] relative to gross national product at current market prices.

	1974	1975	1976	1977	1978	1979
GNP	269 695	298 597	336 374	391 016	473 629	556 779
Agricultural GNP	84 735	94 064	104 657	110 927	131 167	145 616
Research expenditure	202.7	307.0	313.5	384.0	397.7	392.7
As % of GNP	0.08	0.10	0.09	0.10	0.08	0.07
As % of agricultural GNP	0.24	0.33	0.30	0.35	0.31	0.27

[a] 1 B = U.S.$0.05.

agricultural research and during 1974–79 varied from B 1 million to B 13 million. Contributions from private funds were even less meaningful.

Agricultural Research Expenditures by Research Institutions

MOAC and universities are the principal agencies conducting agricultural research and they receive the majority of research funds. During 1974–79, on average, 95% of the total research expenditure went to MOAC, 3.6% to universities, 0.8% to the Ministry of Industry, and 0.6% to the Ministry of Finance (Table 3).

Manpower in Agricultural Research

Agricultural research manpower is controlled by the Civil Service Commission (CSC) under the Prime Minister's Office. It is difficult to ascertain the clear direction and criteria adopted in the past to determine the adequacy of manpower for each research institution.

Total manpower engaged in some form of agriculture-related research activities was estimated to be over 4000 persons in 1979. About 3000 were in MOAC and universities and the remainder were in other ministries. However, based on the criteria for distinguishing research from nonresearch, about one-half of the people were disqualified as research scientists performing research work. The information was inadequate to relate available manpower to the research function of the subunits of all research departments. It was not, therefore, possible to identify the total manpower in the research posts of all research institutes and to determine the total manpower available for research work. The analysis was possible only for MOAC and universities, which constitute the major share of research expenditures and research scientists.

Total agricultural researchers engaged in agricultural research in MOAC and the universities increased from 622 man-year equivalents in 1974 to 2160 in 1979 (Table 4). However, between 1976 and 1979 there was obvious stagnation in manpower as well as in expenditure. This was due to a reorganization of the Department of Agriculture in 1972 that resulted in the separation of extension activities from the department. Assigning staff to various new positions brought about a delay in research project formulation and implementation. The increase in expenditures in 1976 was mainly due to office equipment and salaries. During 1976–79, the Department of Agriculture, which is the main research institute for crop production and usually receives the major share of research expenditure, formulated very few new large-scale research projects.

During 1974–79, the average ratio of Ph.D.: M.S.:B.S.:others was 1:4:36:11, indicating a relatively high concentration of scientists at the B.S. level. Because most research projects are headed by

Table 3. Agricultural research expenditure by research institution.

	1974	1975	1976	1977	1978	1979
Ministry of Agriculture and Agricultural Cooperatives	193.7	295.7	296.8	362.4	377.0	372.5
Universities	5.0	6.8	12.8	16.9	15.7	15.0
Ministry of Industry	2.5	2.5	2.5	2.5	2.5	2.5
Ministry of Finance	1.5	2.0	1.9	2.2	2.5	2.7

Table 4. Total agricultural researchers classified by qualification: numbers are man-year equivalents; figures in parentheses are ratios for each year.

	1974	1975	1976	1977	1978	1979
Ph.D.	8.6	8.6	49.8	48.4	47.4	50.2
	(1)	(1)	(1)	(1)	(1)	(1)
M.S.	32.1	32.5	195.7	186.5	192.6	192.4
	(3.7)	(3.8)	(3.9)	(3.9)	(4.1)	(3.8)
B.S.	532.9	554.1	1034.2	1085.3	1101.0	1187.4
	(62.0)	(64.4)	(20.8)	(22.4)	(23.2)	(23.7)
Others	48.5	49.1	620.1	650.4	670.2	730.3
	(5.6)	(5.8)	(12.5)	(13.4)	(14.1)	(14.5)
Total	622.1	644.7	1899.8	1970.5	2011.2	2160.2

a Ph.D. or M.S., it can be concluded that the research scientists are, on average, adequately supported by technicians.

During the period under study, small increases in manpower were seen only in the B.S. and the lower levels. This is in line with the staffing pattern in the public organizations where government positions are more numerous at the lower levels.

In 1979, of the total manpower available for research work, about 77% of their time was actually engaged in research activities.

Researchers in Research Institutions

The manpower engaged in research work in the universities was 23% of those in the MOAC in 1974 and decreased to 19% in 1979 (Table 5). Prior to 1976 when MOAC was undergoing reorganization, there were more scientists at the Ph.D. level in the university than in the MOAC. From 1976, research manpower at all levels has been higher in the MOAC than in the universities.

Research Expenditures per Scientist

The average research expenditure per scientist man-year during 1974–79 was B 0.26 million (Table 6) compared with B 0.21 million for South and Southeast Asia and B 0.19 million in Asia. However, because of high increases in manpower since 1976 without corresponding increases in the budget, there has been a significant reduction in expenditure per scientist. In addition, the decline in expenditure per scientist has been accompanied by salary increases for the scientists, which has caused the operating funds for each scientist to decline at an alarming rate. Salary increases in the MOAC have gone from 25% of the total expenditures in 1974 to 45% in 1979. The problem in declining operating funds per scientist did not occur in the universities.

Universities were faced with the problem of a relatively small amount of funds per scientist. Scientists in the universities received only 45% of the funds available in the MOAC.

Agricultural Research Expenditures and Manpower in Various Research Disciplines

About 70.3% of the research expenditure was allocated to "multidisciplinary research," 8.9% to plant protection, 8.3% to soil improvement, 4.6% to general cultural practices, 3.3% for agricultural engineering, and the other activities received less than 2% each (Table 7). "Multidisciplinary" research refers to the research programs that consisted of more than one of the other disciplines. These programs were not originally designed as comprehensive multidisciplinary research. Research institutes are organized on the basis of specific commodities and supporting services such as plant protection and soil fertility improvement are handled by other management authorities.

Weaknesses have been indicated in varietal improvement, which could significantly contribute to the agricultural development of the Kingdom. Too few new crop varieties and animal breeds have been introduced to the farmers. No investment has been made in postharvest loss research and, most important of all, physical and biological research programs have not been subjected to economic interpretation.

Of the total man-years engaged in agricultural research, 54.8% were involved in multidisciplinary research, 21.3% in plant protection, 9.4% in cultural practices, 6.6% in soil improvement, and less than 3% in each of the other disciplines (Table 7). The allocation of scientists to various research disciplines was similar to the pattern for the allocation of expenditures and also indicated a weakness in varietal improvement research. However, there was increasing emphasis on the combination of various

Table 5. Distribution of agricultural researchers (man-year equivalents) by research institution and by qualification.

	1974	1975	1976	1977	1978	1979
Ministry of Agriculture and Cooperatives						
Ph.D.	3.2	2.2	43.3	37.8	31.3	34.4
M.S.	28.6	27.5	185.4	167.9	173.3	171.9
B.S.	432.0	400.1	923.3	832.8	841.0	917.4
Others	40.5	40.0	569.6	600.4	620.2	680.0
Subtotal	504.3	469.8	1721.6	1638.9	1665.8	1803.7
Universities						
Ph.D.	5.4	6.4	6.5	10.5	16.1	15.8
M.S.	3.5	5.0	10.3	18.6	19.3	20.5
B.S.	100.9	154.0	110.9	252.5	260.0	270.0
Others	8.0	9.6	50.6	50.0	50.0	50.2
Subtotal	117.8	175.0	178.3	331.6	345.4	356.5
Universities as % of Ministry						
Ph.D.	168	290	15	27	51	45
M.S.	12	18	5	11	11	11
B.S.	23	38	12	30	30	29
Others	19	24	8	8	8	7
Total	23	37	10	20	20	19

research disciplines under the same projects, which is believed to reduce duplication of research effort.

The percentage distribution of research scientists to various research disciplines can be summarized as follows:

	Ph.D.	M.S.	B.S.	Others
Multidisciplinary	49.5	46.6	46.4	73.2
Plant protection	33.5	34.3	22.0	15.1
Soil improvement	3.7	9.0	9.2	1.0
Cultural practices	0.8	0.6	14.8	1.8

The distribution of scientists of different qualifications within each research discipline is presented in Table 8. For each Ph.D. in the various research disciplines, the combination of scientists at the other educational levels varied widely. Research support staff at the B.S. and lower levels were somewhat limited in irrigation, agroindustries, plant protection, and varietal improvement.

Relation of Research Expenditures and Manpower to the Production Value of Agricultural Subsectors

About 86.5% of the total agricultural research expenditures was spent on crop research, 5% on livestock, 4.4% on fisheries, and 0.3% on forestry (Table 9). During the same period, crop production contributed 74% of agricultural GNP, livestock and fisheries about 10% each, and forestry about 6%. Research expenditures were too heavily concentrated on the crop subsector in relation to its production value. On average, the research expenditures relative to the value of production were: crops 0.34%, livestock 0.14%, fisheries 0.15%, and forestry 0.29%.

From 1974–1979 there was no significant change in the research investment on crop production research in terms of its share in the total expenditures and its relation to the value of production (Table 9). Research expenditures for livestock have increased with time. Expenditures for fisheries research as a percentage of the total research expenditure were lowest in 1976 and 1977 relative to other years. Forestry research expenditures have showed signs of improvement recently.

Most of the researcher man-years were engaged in crop production research (84.1% of total), followed by livestock (9.7%), fisheries (3.6%), and forestry (2.6%). Similarly, all levels of scientists were largest in the crop subsector. Eighty-one percent of those with a Ph.D. were in crop production research, 15% in livestock, and the remaining 4% performed research in fisheries and forestry.

Relation of Research Expenditures and Manpower to the Production Value of Individual Agricultural Commodities

The distribution of research expenditures in the crop subsector shows that paddy received an annual

Table 6. Investment (million baht per man-year) in research.

1974	1975	1976	1977	1978	1979	Average
Ministry of Agriculture and Cooperatives						
0.38	0.63	0.17	0.22	0.23	0.21	0.31
Universities						
0.04	0.04	0.07	0.05	0.05	0.04	0.05
Average for Kingdom						
0.33	0.48	0.17	0.19	0.20	0.18	0.26

average between 1974 and 1979 of 20.2% of the total expenditure (Table 10). Multicrop research, which involved more than one crop growing at the same time on different plots of land and was not cropping system research, was allocated 16.7% of the total expenditure. This multicrop research is categorized in this way because it was not possible to obtain reliable information on the crops actually involved. Apart from paddy, rubber, maize/sorghum, vegetables, cotton, and fibre crops received a relatively high share of total research expenditures. Tobacco, coconut, fruit crops, oil crops, and sugarcane/cassava were the crops with a relatively low level of investment in research. Cropping system research received little support and the programs have been operating mainly with assistance from external donor agencies. Cassava, which is important because of the large number of poor farmers involved, did not get financial support. Cassava and sugarcane were grouped together because of the procedure in national accounting for estimating the value of production.

Cotton and coconut are import substitution crops and attempts have been made to improve production. In terms of production values of the crops, the research expenditures allocated to these crops seem relatively large. However, the research results have contributed to some extent to import savings.

Allocations of research expenditures in Thailand bear no relationship to the quantitative importance of individual commodities. This conclusion can be partially explained by the research policy guidelines, which tend to emphasize local needs and local problems rather than specific crops. Even so, some commodities, such as cassava, sugarcane and maize, have been given top priorities as far as agricultural research is concerned. The research emphasis should have been on farming systems and on these crops, but farming systems, cassava, and sugarcane have received relatively little support. The main constraint tends to be the weak link between policy formulation and researchers in the various research institutions, particularly at the departmental level where research projects are normally originated.

Table 7. Agricultural research expenditure (millions of baht) and manpower (man-year equivalents) in various research disciplines. Figures in parentheses are percentage of totals.

	1974		1975		1976		1977		1978		1979	
	Exp[a]	MP	Exp	MP	Exp	MP	Exp	MP	Exp	MP	Exp	MP
Soil improvement	16.0	47.7	22.7	53.1	33.6	126.6	38.4	134.2	39.0	146.5	27.9	110.3
	(7.7)	(7.7)	(7.1)	(8.2)	(9.1)	(6.7)	(9.5)	(6.8)	(9.4)	(7.3)	(6.8)	(5.1)
Irrigation	4.7	30.3	5.5	30.3	8.5	13.6	7.6	13.6	8.0	13.6	—	13.6
	(2.2)	(4.9)	(1.7)	(4.7)	(2.3)	(0.7)	(1.9)	(0.7)	(1.9)	(0.7)	(—)	(0.6)
Engineering	5.9	30.1	12.9	30.1	13.3	37.8	11.6	40.6	11.4	40.4	14.3	42.4
	(2.8)	(4.8)	(4.1)	(4.0)	(3.4)	(2.0)	(2.9)	(2.1)	(2.8)	(2.0)	(3.5)	(2.0)
Toxicology	4.0	20.0	4.8	20.0	6.9	48.7	6.7	49.4	6.3	49.5	8.3	49.5
	(1.9)	(3.2)	(1.5)	(3.0)	(1.9)	(2.6)	(1.7)	(2.5)	(1.5)	(2.5)	(2.0)	(2.3)
Agroindustries	0.1	0.1	0.3	1.3	1.9	3.1	1.2	1.6	0.9	1.9	0.7	1.1
	(0.1)	(0)	(0.1)	(0.2)	(0.5)	(0.2)	(0.3)	(0.1)	(0.2)	(0.1)	(0.2)	(0.1)
Plant protection	18.8	92.4	29.3	101.9	31.7	406.6	33.4	432.2	36.5	442.9	38.6	508.6
	(9.0)	(14.9)	(9.2)	(15.8)	(8.6)	(21.4)	(8.3)	(21.9)	(8.8)	(22.0)	(9.4)	(23.5)
Varietal improvement	3.2	19.0	2.0	22.8	2.9	22.7	3.7	27.6	5.8	43.3	5.5	43.6
	(1.5)	(3.1)	(0.6)	(3.5)	(0.8)	(1.2)	(0.9)	(1.4)	(1.4)	(2.2)	(1.3)	(2.0)
Cultural practices	8.9	130.8	11.9	137.6	13.2	144.8	16.7	138.8	24.9	140.2	21.8	180.3
	(4.3)	(21.0)	(3.7)	(21.3)	(3.6)	(7.6)	(4.2)	(7.0)	(6.0)	(7.0)	(5.3)	(8.3)
Multidisciplinary	147.2	251.6	229.9	247.5	257.5	1096.1	283.1	1132.5	281.4	1131.9	292.9	1210.8
	(70.5)	(40.4)	(72.0)	(38.4)	(69.8)	(57.6)	(70.3)	(57.5)	(68.0)	(56.2)	(71.5)	(56.1)

[a]Exp = expenditure; MP = manpower.

Table 8. Ratio of agricultural research workers with different qualifications in the different research disciplines (average for 1974–79).

	Ph.D.	M.S.	B.S.	Others
Soil improvements	1	9.6	65.1	3.6
Irrigation	1	2.4	2.5	1.2
Agricultural engineering	0	1.1	16.8	19.0
Toxicology	1	3.9	14.1	17.0
Agroindustries	1	1.3	0.3	0
Plant protection	1	4.0	16.9	5.9
Varietal improvement	1	2.0	146.5	0
Cultural practices	1	3.0	453.3	27.7
Multidisciplinary	1	3.7	24.2	19.2

The level of education of the researchers engaged in research work on the different crops was also examined. Crops with a relatively high participation from Ph.D. and M.S. researchers were maize/sorghum, paddy, vegetable crops, cotton, and oil crops. The other crops involved fewer highly educated scientists, but, except for coconuts, all crop research involved some scientists at the level of Ph.D.

Some Problems in the Collection and Interpretation of Data

The fundamental purpose of this study was to contribute to the data base so that further investigations could be conducted on agricultural research, which is an essential element in the development process. A prerequisite for sound policy analysis is an adequate data base. At present, data on agriculture are collected by many different agencies, there is little or no coordination between them, and there is little overall direction on the priorities for data collection and the methods to be employed.

The data presented here may be subject to error. A particular concern is the estimation of manpower engaged in research. Records are poorly kept and not readily available. There is also a possibility of some double counting because competent researchers were involved in more than one project at the same time. Some projects also included nonresearch activities. In these cases, expenditures for nonresearch activities were estimated and excluded from the projects. As well, because there are many research institutes in Thailand and several departments operate research stations situated in various provinces, collecting information from these stations is tedious and time consuming.

Concluding Remarks

A typical farmer in Thailand manages a complex farming system. Commodities are produced for various purposes: some for family consumption, others for cash sale. Most research programs are designed to improve a single crop or animal species rather than to take the farm as an economic unit. As a

Table 9. Research expenditure (millions of baht) relative to value of production (at current market prices) in different agricultural subsectors.

	1974	1975	1976	1977	1978	1979
Crops						
Value	62229	69666	77509	79069	99342	109082
Expenditure	174.2	255.2	273.1	337.6	339.5	333.4
% of value	0.28	0.36	0.35	0.42	0.34	0.31
% of total expenditure	86.4	83.8	89.1	88.5	85.9	95.5
Livestock						
Value	10583	11473	12354	14409	12724	16860
Expenditure	8.2	13.6	19.8	21.3	22.6	23.7
% of value	0.08	0.12	0.16	0.15	0.18	0.14
% of total expenditure	3.9	4.3	5.4	5.3	5.5	5.8
Fisheries						
Value	7273	8454	9792	12456	14103	14584
Expenditure	11.8	27.2	7.8	10.2	17.9	14.1
% of value	0.16	0.32	0.08	0.08	0.13	0.10
% of total expenditure	5.6	8.5	2.0	2.5	4.3	3.4
Forestry						
Value	4650	4470	5002	4995	4998	5090
Expenditure	8.5	11.0	12.8	14.9	17.7	21.5
% of value	0.18	0.25	0.26	0.30	0.35	0.42
% of total expenditure	4.1	3.4	3.5	3.7	4.3	5.3

Table 10. Agricultural research expenditure (millions of baht) on different crops compared with their value of production at current market prices.

	1974	1975	1976	1977	1978	1979	Average
Paddy							
Value of production (1000)	27.9	28.3	25.7	30.2	37.7	40.6	—
Research expenditure	36.3	57.8	60.3	60.2	60.3	65.9	—
% of value	0.13	0.20	0.24	0.20	0.16	0.16	0.18
% of crop research	20.8	22.6	22.1	17.8	17.8	19.8	20.2
Rubber							
Value of production (1000)	3.0	2.3	3.7	4.4	5.6	7.2	—
Research expenditure	21.1	26.7	33.8	34.2	36.0	38.6	—
% of value	0.70	1.17	0.91	0.78	0.65	0.54	0.79
% of crop research	12.1	10.5	12.4	10.1	10.6	11.6	11.2
Sugarcane/Cassava							
Value of production (1000)	6.3	7.9	11.8	9.8	13.1	13.1	—
Research expenditure	7.5	10.2	13.1	15.6	16.1	18.0	—
% of value	0.12	0.13	0.11	0.16	0.12	0.14	0.13
% of crop research	4.3	4.0	4.8	4.6	4.7	5.4	4.6
Maize/Sorghum							
Value of production (1000)	5.9	6.0	4.8	2.3	4.6	6.6	—
Research expenditure	21.7	27.0	30.6	33.2	35.9	35.2	—
% of value	0.37	0.45	0.64	1.44	0.79	0.53	0.70
% of crop research	12.5	10.6	11.2	9.8	10.6	10.6	10.9
Fruit Crops							
Value of production (1000)	7.5	8.4	11.7	13.3	14.7	16.9	—
Research expenditure	2.6	4.0	5.8	7.3	8.2	8.9	—
% of value	0.03	0.05	0.05	0.05	0.06	0.05	0.05
% of crop research	1.5	1.6	2.1	2.2	2.4	2.7	2.1
Vegetables							
Value of production (1000)	4.7	7.6	10.6	9.1	11.4	11.5	—
Research expenditure	14.9	27.1	33.1	34.9	33.7	29.5	—
% of value	0.32	0.36	0.31	0.38	0.30	0.26	0.32
% of crop research	8.6	10.6	12.1	10.3	9.9	8.8	10.1
Oil Crops[a]							
Value of production (1000)	2.9	3.3	3.9	3.8	4.7	5.3	—
Research expenditure	4.1	5.8	7.9	7.5	8.1	8.7	—
% of value	0.14	0.18	0.20	0.20	0.17	0.17	0.18
% of crop research	2.4	2.3	2.9	2.2	2.4	2.6	2.5
Fibre Crops							
Value of production (1000)	1.2	0.9	0.9	1.3	1.4	1.3	—
Research expenditure	9.7	12.8	15.0	25.7	23.41	22.36	—
% of value	0.84	1.34	1.59	1.97	1.63	1.77	1.22
% of crop research	5.6	5.0	5.5	7.6	6.9	6.7	6.2
Cotton							
Value of production	0.4	0.3	0.3	0.7	0.7	0.6	—
Research expenditure	19.5	25.4	28.4	29.2	28.7	30.01	—
% of value	4.55	9.73	9.50	4.40	4.29	4.98	6.24
% of crop research	11.2	10.0	10.4	8.6	8.5	9.0	9.6
Coconut							
Value of production (1000)	0.7	0.5	0.4	0.4	0.7	0.8	—
Research expenditure	3.0	3.6	7.2	7.1	6.6	5.8	—
% of value	0.41	0.78	1.63	1.58	0.93	0.69	1.0
% of crop research	1.7	1.4	2.6	2.1	1.9	1.7	1.9
Tobacco							
Value of production (1000)	1.4	3.8	3.2	3.2	4.1	4.3	—
Research expenditure	1.5	2.0	1.9	2.2	2.5	2.7	—
% of value	0.11	0.05	0.06	0.07	0.06	0.06	0.07
% of crop research	0.9	0.8	0.7	0.7	0.7	0.8	0.8

(Continued)

Table 10. (concluded)

	1974	1975	1976	1977	1978	1979	Average
Other Crops							
Value of production (1000)	0.3	0.4	0.5	0.6	0.6	0.9	—
Research expenditure	2.4	7.1	5.6	7.4	4.7	0.4	—
% of value	0.80	1.78	1.12	1.23	0.78	0.04	0.96
% of crop research	1.4	2.8	2.1	2.2	1.4	0.1	1.7
Multicrops							
Research expenditure	29.2	45.0	22.7	65.2	67.6	60.9	—
% of crop research	16.8	17.6	8.3	19.3	19.9	18.3	16.7
Cropping Systems							
Research expenditure	0.7	0.7	7.7	7.9	7.7	6.42	—
% of crop research	0.4	0.3	2.8	2.3	2.3	1.9	1.7
Total crop research							
expenditure	174.2	255.2	273.1	337.6	339.5	333.4	—

[a]Includes groundnut, mungbean, castor bean, soybean, sesame, and sunflower.
[b]Includes kenaf, ramie, and kapok, but principally kenaf.

result, the small farmer generally is left behind in the technological development process.

Agricultural research, though important, faces several problems. Some of the most important weaknesses include the lack of formal links and proper coordination between several agencies, the absence of an effective mechanism for assigning priorities for research, underfinancing of agricultural research, insufficient relevance of research work to practical farm problems, inadequate collaboration with international research institutes, and inappropriate allocation of research expenditures and manpower to research activities and commodities.

To improve agricultural research in the Kingdom it is recommended that: the availability and utilization of existing research scientists at various research institutes be further assessed and recommendations be made for research manpower management; the recording system for agricultural research projects be improved so that their progress can be monitored and their achievements evaluated; criteria be developed for assigning priorities for agricultural research work; and consideration be given to conducting a case study on the economics of agricultural research work.

Agricultural Research Resource Allocation in Nepal

Ramesh P. Sharma[1]

The objectives of this study were: to describe existing mechanisms for allocating research resources; to prepare an inventory of total agricultural research resources and their distribution over various commodities; and to assess the allocation pattern of resources relative to the importance of commodities in the country. The study is limited to a 5-year period from 1975/76 to 1979/80.

Nepal is predominantly an agricultural country, and agriculture has received top priority in successive plans, as indicated by the allocation of funds to this sector. In spite of considerable investments made in production-augmenting inputs and services during the last decade, the average annual growth rate of agricultural production has been less than 1%. Because the population growth rate is more than 2%, a rapid increase in food production has been the main objective of agricultural development in Nepal.

Agricultural Research in Nepal

The agricultural research system in Nepal was initiated three decades ago, however, research infrastructures have only been established since the early sixties. Agricultural research is conducted by national public institutions. A list of all research institutions and their respective areas of research is given in Table 1, which shows that 15 institutions under four ministries are involved in agricultural research. Due to data constraints, not all research areas could be included in this study. The bulk of agricultural research is conducted within the Ministry of Food and Agriculture. Experimental research farms are scattered in various parts of the country.

There is some amount of overlapping and duplication in research efforts both within institutions and among institutions. Furthermore, as there was no plan to guide establishment of experimental farms in different locations, duplication of work in similar ecological regions is common. A major factor behind this wastage of effort is the lack of an effective institution to decide on priorities and to coordinate research programs in the country. A proposal for creating an Agricultural Research Council has been submitted to His Majesty's Government because it appears that such an institution is urgently needed to guide the agricultural research system in Nepal.

The most important declared objective of agricultural development in Nepal is to increase food production. Agricultural research is expected to play a crucial role by promoting timely modernization, diversification, and continuous improvement. Research that contributes toward increasing production and productivity is to receive top priority.

Resource Allocation System

Financial resource allocation in agricultural research is not different from the budget allocation system of His Majesty's Government of Nepal. The steps followed in budget allocation to research units are: (1) research programs are identified at the national level and allocated to research farms; (2) based on the programs, the research farms prepare an annual budget and submit requests to their department; (3) the department reviews the request and forwards it to its ministry; (4) the ministry reviews and forwards it to the Finance Ministry; (5) the Finance Ministry makes a final review and necessary adjustments; and (6) a sequence of backward readjustments starting from the concerned ministry takes place until the budget at the lowest unit in the hierarchy is ascertained.

Thus the budget adjustment, which in most cases means deduction, occurs at several places. The Finance Ministry is more concerned with the aggregate

[1] Agricultural Projects Services Centre (APROSC), P.O. Box 1440, Lazimpat, Kathmandu.

Table 1. Institutions conducting agricultural research in Nepal (research areas are given in parentheses).

Ministry of Food and Agriculture
Department of Agriculture (Agricultural Crops, Horticulture, Fisheries)
Department of Livestock and Animal Health (Livestock)
Tea Development Corporation (Tea)
Food and Agriculture Marketing Services Department (Agroeconomics, Statistics, Marketing)
Agricultural Projects Services Centre (Agroeconomics)

Ministry of Industry and Commerce
Jute Development Corporation (Jute)
Tobacco Development Board (Tobacco)
Agricultural Tools Factory (Agricultural Tools)
Agricultural Lime Industry (Agricultural Lime)

Ministry of Forest
Botany Department, Department of Medicinal Herbs (Forest Products, Medicinal Herbs)
Forestry Survey and Research Department (Forestry)
Department of Soil and Water Conservation (Soil and Water)

Ministry of Education (University)
Centre for Economic Development and Administration (Agroeconomics)
Agricultural Campus (Agriculture Livestock)
Research Centre for Applied Science and Technology (Agricultural Technology)

---

ministerial budget; whereas, allocation to various research farms takes place mostly at the departmental level. At this level, research must compete with other areas like extension, as well as with various other programs. One important basis of budget allocation is the soundness of the research project. The fact that budget deduction is a very common feature indicates a failure on the part of individual research farms to produce sound research projects that strongly justify the budget request. It is likely that in such a situation informal procedures are more important.

It is suggested that for improvement either: (1) there should be an effective organization to bargain for research programs; or (2) the budget allocation process for research should be made more independent of the governmental budget allocation system.

Because agricultural research is entirely in the public sector, manpower allocation is the same as in other public sectors. A departmental request for a new post must be submitted to the Finance Ministry and the Administrative Management Department (AMD), the former for budget provision and the latter for the sanction of the post. A committee of representatives from these two organizations and the concerned ministry, reviews the case and forwards it to Cabinet. Once the Cabinet approves the post, the Public Service Commission advertises, selects, and recruits the staff. The whole process can take 6 months to more than 1 year. To a large extent the time required for the whole process depends upon how matters are expedited, both formally and informally. Recruitment to a new permanent post is a lengthy and cumbersome process.

Temporary recruitment is less stringent. After the approval of the committee, the concerned department can advertise and recruit its staff. But the motivation to work is low because the person will be released as soon as the Public Service Commission manages to select a candidate for the permanent post.

One prompt and efficient means of mitigating a temporary shortage of staff in a particular area is to send someone on deputation. As recruitment of new staff takes time and may be constrained by nonavailability of trained manpower, this system is very popular. However, improper influences are often used to secure transfers to attractive areas at the cost of continuity of involvement in the former post. Research suffers from this discontinuity.

Resource Allocation Trend

Financial Allocation

During 1975/76 to 1978/79[2] the average annual rate of increase in the agricultural research budget was 21%; food crops and agricultural engineering research recording the highest rates of increase. During 1975/76 to 1979/80, of the average total research budget of Rs. 26 million[3] per year, 68% was allocated to crops (about 53% to food crops and 16% to cash crops). When research areas like basic biology, agricultural engineering, and soil/water, which indirectly support crops research, are included, almost 83% of the research budget was allocated for crops. Horticulture and forestry research each claimed about 6% of the total budget, while 4% was spent on livestock and fishery research.

Over the years, the share of the research budget allocated to food crops has increased. It declined for cash crops, horticulture, and fisheries but remained constant in areas like biology, engineering, and soil/water (Tables 2 and 3).

During the study period, food crops received 77% of the crops research budget. Paddy, maize,

---

[2] Due to some exogenous reason that applied to all governmental programs, the budgets allocated to most sectors declined in 1979/80 relative to 1978/79.

[3] Rs. 12 = U.S.$1.

Table 2. Commodity-oriented agricultural research budget allocation (in thousands of rupees).

	1975/76	1976/77	1977/78	1978/79	1979/80
Cereal Crops	*8110*	*12470*	*14446*	*19042*	*14243*
Paddy	2374	4749	4879	6459	4699
Maize	3006	3770	4635	6428	5079
Wheat	2439	3656	4620	5770	4113
Millet	30	35	55	67	70
Barley	104	92	100	106	113
Pulses	48	68	71	95	103
Others	73	100	86	117	66
Cash Crops	*3970*	*3389*	*3413*	*4218*	*5342*
Potato	391	659	878	1021	2587
Sugarcane	959	509	441	527	435
Oilseeds	1062	835	454	808	505
Cotton	536	300	356	382	320
Jute	315	360	495	640	690
Tobacco	380	420	480	630	620
Cardamom	109	159	136	193	154
Ginger	218	147	173	17	31
Tea[a]	—	—	—	—	—
Livestock	549	658	738	757	795
Horticulture	1685	1516	1718	1879	1514
Fishery	291	530	382	381	290
Forestry	1088	1656	1466	1901	2109

[a]There are no research programs on tea.

Table 3. Budget allocation (in thousands of rupees) to research activities that are difficult to attribute to commodity categories.

	1975/76	1976/77	1977/78	1978/79	1979/80
Agri. botany	525	448	451	500	538
Plant pathology	484	500	564	600	542
Entomology	519	525	536	1187	739
Soil science and agri. chemistry	613	715	897	861	864
Agri. engineering	530	678	807	1245	1064
Soil and water resources	350	437	386	433	481

and wheat received 97.6% of the total food crops; whereas, only 2.4% was spent for research on millet, barley, and pulses. Cash crops research claimed 16% of the total agricultural research budget and 23% of the crops research budget. Of the total cash crops budget, potatoes received the largest share (27%) followed by oilseeds (18%), sugarcane (14%), tobacco and jute (each 12%), cotton (9%), cardamon (4%), and ginger (3%). Tea is an important cash crop for Nepal, but no research program has been initiated.

Manpower Allocation

Total agricultural research manpower (including nonresearch areas) increased from 352 (officers) and 1045 (assistants)[4] in 1970/71 to 773 (officers) and 2450 (assistants) in 1979/80. In terms of administrative division of officer level manpower, 7% are in

[4] Officer level: those with B.Sc. or above; assistant level: matriculation and 1–2 years training.

Class I, 22% in Class II, and 71% in Class III. Similarly, 2% have a Ph.D., 26% have a M.Sc., and 72% have a B.Sc. The vacancy rate is about 15% at the officer level and 9% at the assistant level. Of the total agricultural manpower, 29% at the officer level and 12% at the assistant level are engaged in research.

The total number of agricultural research scientists in Nepal is 226, which means that there are 17 research scientists for every 1 million people in the agricultural population. The distribution of research manpower in different research areas is shown in Tables 4 and 5. The highest concentration of manpower is in crops research, about 47% of all manpower. However, when noncommodity research areas indirectly supporting crops research are taken into account, crops research engages 80% of all research manpower. After crops, research on forestry, horticulture, livestock, and fisheries have received the remaining manpower (Table 4).

About 77% of all officer level crops research manpower is allocated to food crops, the rest goes to cash crops. Paddy, maize, and wheat have together claimed 95% of all food crops research manpower and 67% of all crops research manpower. Millet, pulses, and barley research programs received only 5% of all manpower engaged in research on food crops.

In cash crops, cotton research has engaged 35% of cash crops research manpower. Of the remainder, 19% are in the sugarcane program, 13% each in the potato, oilseeds, and tobacco programs, and the rest are in jute and cardamon.

Assessment of Resource Allocation Pattern

Financial Resource

The average agricultural research expenditure between 1975/76 and 1979/80 was 0.15% of GDP and 0.23% of the agricultural GDP. Research expenditure relative to agricultural GDP appears to be somewhat higher than in countries like Indonesia and Bangladesh but considerably lower than in developed countries and some other Asian countries.

Research investment relative to the value of production is: 0.33% in crops, 0.24% in horticulture, 0.02% in livestock, 0.26% in fisheries, and 0.21% in forestry. Thus, investment in crops research in relative terms is higher than in other sectors. Relative investment in livestock research is very low (Table 6).

The annual rate of growth in research investment relative to value of production is highest in crops, constant in livestock, and fluctuates in horticulture,

Table 4. Distribution of manpower (in man-years) involved in agricultural research by commodity (1980).

	Scientists[a]	Assistants[b]
All Crops	*97.3*	*149.2*
Food Crops	*75*	*96*
Paddy	27	35
Maize	24	28
Wheat	20	29
Millet	0.5	1.5
Barley	1.1	0.7
Pulses	2.1	1.7
Cash Crops	*22.3*	*53.2*
Potato	3	5.4
Sugarcane	4.3	6.4
Oilseeds	3	5.4
Cotton	7	21
Jute	1	4
Tobacco	3	7
Cardamom	1	4
Horticulture	13.5	18
Livestock	10	6
Fishery	4	6
Forestry	21	22

[a]Those with the degree of B.Sc. or above.

[b]Junior technicians (JT) have matriculation and 2 years training; junior technical assistants (JTA) have matriculation and 1 year training.

Table 5. Distribution of manpower (in man-years) in research areas not attributable to commodities (1980).

Research area	Scientists	Assistants
Entomology	15	14
Soil science and agri. chemistry	19	18
Agri. botany	12	15
Plant pathology	15	13
Plant quarantine	6	8
Agronomy	3	1
Agri. engineering	8	13
Soil and water research	2	5

fisheries, and forestry. This shows that unless corrective steps are taken, only crops research investment is likely to move in accordance to its value of production.

The distribution of research investment as a percentage of production value for the various crops during 1975/76 to 1979/80 was: millet and pulses, 0.01–0.02%; paddy and barley, 0.19%; oilseeds and jute, 0.24–0.29%; maize and wheat, 0.42–0.49%; and sugarcane and tobacco, 0.60–0.67%. The value

Table 6. Agricultural research expenditure relative to agricultural GDP and total agricultural research expenditure (figures in millions of rupees).

	1975/76	1976/77	1977/78	1978/79	1979/80	Average
Total Agri.						
Res. Expend.						
(TARE)	17.778	22.816	25.053	32.295	27.950	25.178
Crops						
Agri. GDP	6746	6121	6168	6411	5743	6238
Res. expend.[a]	14.165	18.456	20.749	27.377	23.242	20.798
As % of agri. GDP	0.21	0.30	0.34	0.43	0.40	0.33
As % of TARE	79.7	80.9	82.8	84.8	83.2	82.6
Horticulture						
Agri. GDP	699	668	668	711	721	693
Res. expend.	1.685	1.516	1.718	1.879	1.514	1.662
As % of agri. GDP	0.24	0.23	0.26	0.26	0.21	0.24
As % of TARE	9.5	6.6	6.9	5.8	5.4	6.6
Livestock						
Agri. GDP	3482	3324	3358	3394	3421	3396
Res. expend.	0.549	0.658	0.738	0.757	0.795	0.699
As % of agri. GDP	0.02	0.02	0.02	0.02	0.02	0.02
As % of TARE	3.1	2.9	2.9	2.3	2.8	2.8
Fishery						
Agri. GDP	132	133	134	159	166	145
Res. expend.	0.291	0.530	0.382	0.381	0.290	0.375
As % of agri. GDP	0.22	0.40	0.28	0.24	0.17	0.26
As % of TARE	1.6	2.3	1.5	1.2	1.0	1.5
Forestry						
Agri. GDP	556	895	813	805	882	790
Res. expend.	1.088	1.656	1.466	1.901	2.109	1.644
As % of agri. GDP	0.20	0.18	0.18	0.25	0.24	0.21
As % of TARE	6.1	7.3	5.8	5.9	7.5	6.5

[a]Includes research areas like agricultural engineering, basic biological research, and soil/water research, which indirectly support crops research, but excludes some crops whose contribution to agricultural GDP is not available. They are some minor food crops, cotton, cardamom, and ginger.

for all food crops was 0.26% and for all cash crops was 0.33%. These figures show that all food crops, other than maize and wheat, seem underinvested in research; whereas, on the whole, cash crops research has received a high level of investment relative to its production value.

When the level of investment is compared to cultivated area, food crops appear underinvested (to the extent of 11%) and cash crops overinvested (to the extent of 50%). The millet and paddy research programs are especially underinvested (Table 7).

Relative to the importance of food crops in the consumption pattern in Nepal, all food crops except wheat have been underinvested. Millet and barley research show considerable underinvestment (Table 7).

The three criteria used here broadly confirm the imbalance in relative resource allocation in crops. With reference to these three criteria, rice and millet research is underinvested; maize and wheat research is overinvested.

Manpower Resource

The percentage distribution of agricultural research manpower and the value of production of the various sectors is:

	Crops	Horti-culture	Live-stock	Fish-eries	For-estry
Value	56	6	30	1	7
Manpower	66	9	7	3	14

These figures show a relatively uneven distribution in value of production and manpower (excluding noncommodity research manpower). The livestock sector appears grossly understaffed, whereas fisheries and forestry are relatively overstaffed. Relatively, the crops sector appears to be nearest to the optimal allocation pattern.

46

Table 7. Percentage distribution of research investment relative to area under crops and consumption pattern.

| | Research | | Culti-vated area | Consumption levels |
	Invest-ment	Man-power		
Paddy	27	31	52	42
Maize	27	28	18	36
Wheat	24	23	14	12
Millet	0.3	0.6	5	10[a]
Oilseeds	4	3.2	5	—
Potato	7	3.2	2	—
Others[b]	10	11	4	—

[a]Out of this, 7% is millet and 3% is barley.
[b]Includes barley, sugarcane, tobacco, and jute.

A disproportionate distribution of manpower relative to the value of production of various crops was also observed. Research programs in millet, pulses, and paddy in food crops and potato, oilseeds, and jute in cash crops, appear understaffed. A wide deviation in manpower deployment relative to value is evident in millet, pulses, and jute.

When manpower is compared with the distribution of cultivated area, paddy, millet, and oilseeds seem understaffed. All the remaining crops are relatively overstaffed (Table 7). On the basis of consumption criteria, underinvestment in research manpower is evident in all food crops except wheat. However, the extent of underinvestment is negligible in paddy and maize (Table 7).

Some Other Issues

Agroeconomics Research

Due to a data problem, estimates of resources allocated to agroeconomics research were not included in the earlier analysis. In the year 1979/80 approximately Rs. 3.7 million was spent on agroeconomics research (mostly of a multicommodity nature) by the three public institutions involved in this area (Table 1). The Rs. 3.7 million is about 11% of the total agricultural research budget in 1979/80. It is estimated that there are 53 posts in agroeconomics and agristatistics within agriculture-related institutions, but none exclusively in research.

Integrated Cereals Project

This 5-year project was initiated in 1976 with assistance from USAID. The main components of the project are: (1) a strengthening of the existing research base of cereal crops by logistic and training supports; (2) an initiation of a research program in cropping systems; and (3) an introduction of a minikit program of improved cereal crop varieties, mainly rice, maize, and wheat, and other inputs for wider adoption of modern technology. The most significant aspect of the project is the initiation of a cropping systems research program, although the bulk of the total project budget of U.S.$9 million was spent on advanced level academic training abroad.

Locational Aspect of Research System

Geographical and regional imbalances in the distribution of research farms and stations exists in Nepal. Mountainous and hilly regions, which occupy two-thirds of the total land area, have the least concentration of research facilities and manpower. An exception to this is the Kathmandu Valley. Furthermore, there are many experimental farms where research facilities are poor, research staff very few, and budgets too small to operate effectively.

Future Manpower Scene

During 1980/81 to 1984/85 (Sixth Plan Period) there will be a surplus of professional manpower in agriculture; however, the livestock sector will face a manpower deficit. Ineffective use of existing manpower is considered more of a problem in Nepal than is its shortage. The ratio of expatriate agricultural research manpower to total manpower in Nepal is negligible.

Conclusions

(1) Agricultural research investment in Nepal is low relative to some other countries of Asia. Further investment in research seems necessary. The extent and the nature of further investment should be ascertained on the basis of cost-benefit studies.

(2) Examination of the resource allocation pattern to different subsections revealed an abysmally low level of financial and manpower resources in livestock research relative to its value of production. This serious imbalance should be corrected. Livestock rearing is an integral part of the Nepalese farming system. In general, agricultural development measures in Nepal have shown a bias toward the crops sector, and this is also the case for research.

(3) During the last 5 years, only research investment in the crops sector has been increasing relative to growth in its value of production. Unless investments in other sectors are also increased in proportion to the value of production, the existing misallocation of resources among agricultural subsectors will be prolonged.

(4) The most important objective of agricultural development in Nepal is to increase food grain production. Cash crops development is the next

most important objective. In view of this, relative allocation of resources between food and cash crops assumes importance. This study revealed that research on food crops is relatively underinvested and understaffed compared with cash crops research. Thus, the present allocation pattern contradicts the declared objectives of agricultural development. This misallocation should be corrected.

(5) Further disaggregation of crops showed that research on paddy, millet, and pulses is underinvested and understaffed. Maize and wheat research programs appear to have received larger resources than their importance would justify.

Inadequate resources for paddy research should be considered as a serious imbalance because paddy is the most important crop in Nepal in terms of area, production, value, consumption, and exports. Millet and pulses are important consumption crops. A commodity development program to look after these and other minor crops appears necessary in Nepal.

Maize is a staple crop in the hills and has a large production potential. In view of recurring food shortages in the hills and the suitability of maize farming to hilly terraces, the relatively high level of resources allocation for this crop should continue.

In recent years wheat has become the most important winter crop in Nepal. More than 90% of the total wheat area in Nepal is covered by modern varieties. This may partly be explained by a larger proportion of research resources allocated to wheat research. Although a switch from traditional to modern varieties has been accomplished, the productivity of wheat has not improved. It is suggested that future wheat reseach programs should concentrate mainly on factor research rather than varietal selection. Efficient use of soil, fertilizer, and conservation of soil fertility should be accorded top priority in this factor research.

(6) Cash crops research programs have received larger resources than their importance would justify. Given the fact that the stock of resources at a particular time is limited, transfer of some amount of resources from cash crops research to food crops would be beneficial to the economy.

(7) Initiation of the cropping systems research program in 1976 is a laudable step. This program should receive adequate resources and be continued in the future.

(8) Monitoring, evaluation, and special studies on agricultural research in particular and related agricultural policy issues in general are far from adequate in Nepal. Studies on productivity of agricultural research investment by commodity, continuous monitoring on research resources allocation, research resources and programs for minor crops, and development of an effective evaluation system appear rather urgent.

The Agricultural Research Resource Allocation System in Peninsular Malaysia

Nik Ishak bin Nik Mustapha[1]

Agricultural research in Peninsular Malaysia involves the Ministry of Agriculture and the Ministry of Primary Industries. The agencies involved in agricultural research under the Ministry of Primary Industries are the Rubber Research Institute of Malaysia (RRIM), the Palm Oil Research Institute (PORIM), and the Forestry Research Institute. Under the Ministry of Agriculture, the Malaysian Agricultural Research and Development Institute (MARDI) and the Fisheries Research Institute are engaged in research activities. Other agencies under the Ministry of Agriculture such as the Federal Agricultural Marketing Authority (FAMA), the Bank Pertanian Malaysia (BPM), and a few other agencies are also involved in research; their involvement, however, is negligible and more on an ad hoc basis. The private sector, especially the big corporations, also conducts research. Research by the private sector is mainly for their use and results and financial involvements are not readily available to the public.

Most of the government agencies that conduct research, except for MARDI, are only involved in a single commodity. MARDI, which was established in 1969 and became operational in 1971, is entrusted with the responsibility of conducting research on all agricultural commodities except for rubber, timber, marine fisheries, and lately oil palm (with the formation of PORIM in 1979). It is for this reason that the total expenditure of MARDI cannot be readily allocated to commodities and a resource allocation study is extremely important.

Scope, Objectives, and Methodology

This research resource allocation study, a project approved by the MARDI Board, involves the allocation of MARDI's financial resources to specific commodities. This study does not encompass the allocation of research resources of other agencies. However, some additional data fom other relevant agencies were collected. The time frame for the MARDI study was 1975–79.

The objectives of the MARDI study were to: (1) study the allocation of research resources over the last 5 years according to commodities; (2) investigate discrepancies, if any, between the amount of resources expended on the commodities and their contribution to the national economy; (3) provide a data base for a more rational system of resource allocation in the future; (4) develop a system for monitoring future resource allocation; and (5) provide a data base for future studies on costs and returns to research on major commodities.

A short-term consultant was employed to develop the methodology for this study. Modifications to the initial methodology were done to increase the precision of the data. Generally, the methodology involved the collection of actual expenditure data segregated into expenditure components (codes) and expenditure units. Allocating percentages were constructed for each of the expenditure units by analyzing the disaggregatable portions of the total expenditure that were actually a part of the line project costs (could be identified to commodities). The allocating percentages were then used to disaggregate the total expenditure of each expenditure unit into commodities. The expenditures of each commodity for all the expenditure units were summed to give the total MARDI expenditure for each commodity. It should be cautioned that this method assumes that the

[1] Research Officer, Economic Branch, Malaysian Agricultural Research and Development Institute, Bag Berkunci No. 202, Pejabat Post Universiti Pertanian, Serdang Selangor, Malaysia.

Table 1. Agricultural research expenditure as a percentage of GNP and as a percentage of agricultural component of GDP.

	1973	1974	1975	1976	1977	1978	1979
Total agriculture research expenditure[a] (mill. MR)	42.88	57.48	67.26	53.87	68.40	76.49	80.99
GNP at market prices (mill. MR)	17 963	21 861	21 606	27 033	31 074	35 090	40 740
Agricultural component of GDP[b] at market prices (mill. MR)	5425	6572	6177	7693	8430	8975	10 200
Agricultural research as % GNP	0.24	0.26	0.31	0.20	0.22	0.22	0.20
Agricultural research as % agricultural GDP component	0.79	0.87	1.09	0.70	0.81	0.85	0.79

[a] Total agricultural research expenditure is a sum of MARDI, RRIM, Forestry Research Institute, and Fisheries Research Institute expenditures.

[b] Agriculture component of GDP includes agriculture, livestock, forestry, and fishing.

nondisaggregatable portion of the MARDI expenditure is proportional to the disaggregatable portion. For instance, if station A's (an expenditure unit) disaggregatable expenditure is 70% rice and 30% vegetables, then 70% of the nondisaggregatable expenditure is allocated to rice and the remaining 30% to vegetables.

Agricultural Research Expenditure

The total agricultural research expenditure for Peninsular Malaysia is the sum of the MARDI, RRIM, Forestry Research Institute, and Fisheries Research Institute expenditures (Table 1). Other agencies also conduct agricultural research, but the research expenditure data are not available and are negligible. A major portion of the expenditure is incurred by RRIM and MARDI (about 80–90%). Prior to 1976, the research expenditure for RRIM was significantly higher than that for MARDI. This has been reversed since 1976.

The agricultural research expenditure is only about 0.2–0.3% of GNP. A more meaningful measure is probably the proportion of research expenditure to the agricultural GDP and this is generally slightly less than 1%. Data on the national research and development expenditure are not available.

Expenditure by Commodity

For 1975–79, research on all agricultural commodities except rubber, timber, and most of marine fisheries was under MARDI. In terms of utilization of financial research resources, rubber was by far the most important commodity. Most of the remaining research resources went to commodities such as rice, timber, oil palm, vegetables, cocoa, beef, and fruits (Table 2). It is no surprise that rubber consumes a considerable chunk of the financial research resources. Research on rubber falls under one institute (RRIM), and a minimum fund is required to set up and maintain a research organization. This is also true for forestry research. Rubber and forestry products after all are the major agricultural revenue earner for Malaysia. Research on other agricultural commodities falls under MARDI. From late 1979, a new research organization (PORIM) was formed to conduct research on oil palm, which was formerly done by MARDI. Research resources for oil palm can thus be expected to increase considerably in the near future.

Expenditure in Relation to Production Value

Production value could only be determined for some of the commodities because of a shortage of data. For some commodities such as vegetables and fruits, which can be further broken down to different types, the estimation is difficult because prices and production figures of the different types are unavailable over time. The production value of a few commodities was estimated and the percentage of this value devoted to research was calculated (Table 3).

In the case of rubber, the proportion appears to decline through the years. This is because the re-

Table 2. Agricultural research expenditure (mill. MR) on different commodities by MARDI and by research organizations other than MARDI.

	1976	1977	1978	1979
Other than MARDI				
Rubber	25.10	28.89	26.55	27.42
Timber	3.11	3.81	6.67	4.72
Marine fisheries	0.002	1.28	1.80	1.96
MARDI				
Rice	4.82	6.49	7.40	9.47
Oil palm	3.24	4.98	4.90	2.90
Cocoa	1.22	1.30	2.63	4.19
Coconut	1.00	0.54	0.59	0.90
Sugarcane	0.48	0.62	0.66	0.70
Root crops	0.25	0.45	0.65	0.72
Cereals	0.20	0.11	0.50	0.53
Legumes	0.49	0.17	0.55	0.90
Fruits	1.14	1.53	2.18	2.62
Vegetables	2.12	2.41	2.72	3.74
Tobacco	0.35	0.24	0.38	1.00
Ornamentals	0.06	0.20	0.17	0.10
Pineapple	0.81	1.22	1.90	1.00
Spices	0.25	0.30	0.22	0.53
Beef	2.67	3.22	4.19	4.72
Dairy	0.49	0.53	0.93	0.71
Poultry	0.48	0.54	0.59	0.47
Swine	0.18	0.92	0.76	0.61
Small ruminants	0.12	0.32	0.28	0.34
Freshwater fish	0.66	0.81	0.91	1.93
Soils	0.63	1.31	0.77	1.09
Agri. product utilization	1.15	2.40	3.12	3.05
Engineering/water management	0.05	0.30	0.48	0.18
Pasture	1.56	1.19	0.71	1.24
Feed	0.45	0.67	0.49	0.81
Cropping system	0.35	0.37	1.35	0.28
Others	1.30	1.30	1.56	2.12

search cost was stable while the production value increased. The proportion for oil palm has been constant, except for a sudden drop in 1979. This drop is a result of both an increase in production value and a drop in research expenditure. The drop in expenditure may have been due to the knowledge that oil palm was to be taken out of MARDI. Relative to other commodities, the proportion of research for marine fisheries and timber was small. Research for rice was about 1–2.4% of the production value (except for 1978). The highest proportion was for beef and this may be attributable to the costly nature of beef research. The proportion of research on cocoa has been increasing since 1977, because MARDI has given cocoa higher priority.

Manpower Availability

MARDI personnel can be divided into four levels: group A are those with at least a first degree or its equivalence; group B hold diploma/higher school certificates or their equivalence; group C are those with a minimum qualification school certificate; and the majority of group D are labourers (Table 4). The number of personnel in MARDI increased tremendously from a core of 422 in 1971, to 2323 in 1975, and finally to 3599 in 1980. This represents a 5.5 fold increase from 1971–1975 and a 1.5 fold increase for 1975–80. About 70–80% of the personnel are in groups C and D and the upsurge in 1971–75 was attributed mainly to the increased intake of staff in these categories. Because MARDI is a research organization, the number of technical staff dominates the staff that can be classified as administrators. This is reflected in the number of personnel categorized as professional, subprofessionals, and technical in groups A, B, and C, respectively.

Time-series data for the number of personnel by discipline are not available. However, the proportion is not expected to vary much over time. Table 5 gives the complete breakdown of researchers by discipline in MARDI in 1980. Researchers registered as agronomist comprised about 19% of all researchers.

Other Criteria for Judging Appropriateness of Allocation

The priority that is accorded, or should be accorded, to commodities is a function of a number of variables, only some of which are quantitative. These priorities should be tied to the overall national policy. Variables such as potential to eradicate poverty, creation of employment opportunities, and political importance of the commodity are some nonquantifiable criteria. More often than not, these nonquantifiable criteria can override the quantifiable variables. It would be interesting to conduct a separate study of these nonquantifiable criteria in Malaysia.

In this study, two additional quantitative criteria were included besides the productive value of the commodities: (1) the import value of the commodities, because this may have an important bearing on the import substitution policy; and (2) the hectarage, which may be important in terms of space coverage by the commodities (Table 6). Based on import value alone, for 1979 it appears that rice should be

Table 3. Research expenditure on various commodities as a percentage of production value.

	Rubber	Oil palm	Timber	Marine fisheries	Rice	Beef	Cocoa
1975							
Research (mill. MR)	37.78	3.35	2.87	0.09	4.99	2.76	1.26
Prod. value							
(mill. MR)	2026	1320	1061	532	461	63	42
Res./prod. value (%)	1.86	0.25	0.27	0.02	1.08	4.38	3.03
1976							
Research (mill. MR)	25.10	3.24	3.11	0.002	4.82	2.67	1.22
Prod. value							
(mill. MR)	3117	1216	2325	642	290	59	69
Res./prod. value (%)	0.80	0.26	0.13	0.003	1.66	4.49	1.77
1977							
Research (mill. MR)	28.89	4.98	3.81	1.28	6.49	3.22	1.30
Prod. value							
(mill. MR)	3379	1796	2307	828	267	25	148
Res./prod. value (%)	0.85	0.27	0.16	0.15	2.43	13.11	0.88
1978							
Research (mill. MR)	26.55	4.90	6.67	1.80	7.40	4.19	2.63
Prod. value							
(mill. MR)	3599	1871	2467	1293	148	47	151
Res./prod. value (%)	0.73	0.26	0.27	0.14	4.99	8.94	1.74
1979							
Research (mill. MR)	27.42	2.90	4.77	1.96	9.47	4.72	4.19
Prod. value							
(mill. MR)	4509	2378	2854	988	408	NA	189
Res./prod. value (%)	0.60	0.12	0.16	0.20	2.32	—	2.22

Table 4. Availability of different levels of manpower in MARDI.

	1971	1972	1973	1974	1975	1976	1977	1978	1979	1980
Group A										
Management	4	6	6	6	7	9	11	13	14	18
Professional	69	97	115	149	234	269	284	324	365	386
Group B										
Executives	5	5	4	7	7	7	8	16	22	22
Subprofessional	49	94	88	153	200	273	255	284	331	355
Group C										
Clerical	22	37	39	50	72	83	84	106	134	140
Technical	88	126	196	270	324	373	371	438	555	682
Group D	185	241	326	517	1479	1425	1438	1564	1858	1996
Total	422	606	774	1152	2323	2439	2451	2745	3279	3599

given top priority followed by fruit, vegetable, marine fisheries, and beef. But, it should be noted that both fruits and marine fisheries seem to exhibit a strong upward trend of import value. This has an important bearing on the future allocation of research resources. On the basis of area, rubber should be accorded top priority followed by oil palm, rice, fruits, cocoa, and vegetables.

Discussion and Summary

This country inventory report for Malaysia is based on a research resource allocation study for MARDI and some additional information gathered from other relevant agencies. It is an initial attempt to classify agricultural research resources by commodity. This inventory does not encompass

Table 5. Number of researchers in different disciplines at MARDI in 1980. Source: Register of Researchers (1980), MARDI.

Discipline/Specialization	Number
Agronomist (crops and fertility)	68
Food scientist	33
Plant breeders	27
Soil scientist	24
Economist/Sociologist/Agribusiness	21
Entomologist	20
Chemist/Agri. chemist	13
Agri. engineers/Engineers	12
Animal nutritionist	11
Statisticians/Computer sc./Data analyst	7
Ag. extension/Communication/Journalism	7
Animal husbandry	7
Genetics	7
Other animal scientist	5
Biochemist	5
Postharvest/Processing scientist	5
Fish technology/Breeding/Nutrition	5
Dairy technology/Husbandry	4
Veterinary medicine	4
Horticulturist	4
Agriculture	4
Weed scientist	4
Bacteriologist	3
Mammalogy	3
Meat technology	3
Irrigation and agri. drainage	3
Others	30

agricultural research activities in the private sector but does cover most of the agricultural research activities conducted by government agencies. With regard to manpower resources, this report only covers MARDI.

The current financial practice adopted by MARDI, does not allow us to determine the proportion of the financial resources going to each commodity. Because it is the largest and the only multicommodity agricultural research organization, this information is vital to MARDI for future planning purposes if the money is to be allocated according to the priorities set for each commodity. Information on expenditures for specific commodities would also be useful for benefit-cost studies on specific commodities. The country inventory as a whole can provide useful insights for the expansion or reduction of resources for specific commodities or institutes.

Some misallocation of resources can be identified on the basis of this initial investigation. For instance, MARDI, because it must cover research on more than 20 commodities, needs extra resources. Currently, MARDI is allocated about one-half of the total financial resources. Marine fisheries should also receive more research funding. Too much money appears to have been channeled to beef re-

Table 6. Area of cultivation and import value of selected commodities.

	Rubber	Oil palm	Cocoa	Rice	Vegetables	Fruits	Beef	Marine fish
1975								
Research (mill. MR)	37.78	3.35	1.26	4.99	—	—	—	0.09
Import value (mill. MR)	—	—	—	88	46	27	9	30
Area ('000 ha)	1695	569	17	595	6	75	—	—
1976								
Research (mill. MR)	25.10	3.24	1.22	4.82	2.12	1.14	2.67	0.002
Import value (mill. MR)	—	—	—	75	44	25	14	35
Area ('000 ha)	1701	638	20	580	5	83	—	—
1977								
Research (mill. MR)	28.89	4.98	1.30	6.49	2.41	1.53	3.22	1.28
Import value (mill. MR)	—	—	—	92	58	32	20	36
Area ('000 ha)	1703	712	29	567	6	85	—	—
1978								
Research (mill. MR)	26.55	4.90	2.63	7.40	2.72	2.18	4.19	1.80
Import value (mill. MR)	—	—	—	22	52	48	26	44
Area ('000 ha)	1711	764	26	446	9	86	—	—
1979								
Research (mill. MR)	27.42	2.90	4.19	9.47	3.74	2.62	4.72	1.96
Import value (mill. MR)	—	—	—	85	52	72	25	48
Area ('000 ha)	1727	821	32	562	9	85	—	—

search. As noted earlier, a high value can be attached to nonquantifiable criteria such as the potential of the crop to reduce poverty. This is especially so in the case of rice, but the production value, hectarage, and import value are insufficient to indicate whether enough resources were channeled to this crop. Because MARDI is the largest agricultural research organization and deals with a number of commodities, it is appropriate that it conduct this type of study. I strongly suggest that a separate study to formulate priority indices for Malaysian agricultural commodities be conducted.

Resource Allocation to Agricultural Research in Pakistan

Malik Mushtaq Ahmad[1]

Agriculture is the mainstay of Pakistan's economy. It contributes 29.1% to GNP, employs 52.9% of the total labour force, and provides livelihood to over 70% of the population. Of the total agricultural contribution to the GNP, crops contribute 67.9%, livestock 29.9%, fisheries 1.6%, and forestry 0.7%.

The total area of Pakistan is 79.61 million hectares, of which 20.12 million hectares are cultivated. The major crops grown are wheat, rice, cotton, sugarcane, maize and millets, and gram and pulses, which constitute 35.0%, 10.6%, 9.8%, 3.9%, 9.3%, and 8.7% of the cropped area, respectively. The livestock population consists of 14.85 million cattle, 10.61 million buffalo, 18.93 million sheep, 21.69 million goats, and a total of 3.44 million horses, donkeys, and camels. Annual fish production is 0.3 million tonnes, of which 13.6% is from inland waters and 86.4% is marine. Forests cover 3.5% of the area and produce 229 000 m^3 of timber and 576 000 m^3 of firewood.

Research System

To obtain sustained agricultural production, a well-structured and functional agricultural research system is necessary. At Independence in 1947, Pakistan inherited only one research institute on irrigation and the defunct Punjab Agricultural College and Research Institute, Faisalabad. These institutions were inadequate to support the research and development requirements of the large agriculture sector. However, Pakistan has developed a potentially viable research system.

The research system is composed of a number of diverse research establishments and organizations that are administratively controlled by different federal and provincial ministries. At the federal level, the Pakistan Agricultural Research Council (PARC) operates three institutes and six research units; the Pakistan Atomic Energy Commission (PAEC) maintains three institutes; the Water and Power Development Authority (WAPDA) has one directorate on soil and water quality monitoring; the Pakistan Central Cotton Committee (PCCC) operates three institutes on cotton. In addition to the PCCC, the Federal Agricultural Ministry has under its direct control, the Soil Survey of Pakistan and the Pakistan Forest Institute. The Irrigation Drainage and Flood Control Research Council is under the Ministry of Science and Technology. The Pakistan Tobacco Board is under the Ministry of Commerce. Other institutions, including the universities, are under the administrative control of the respective provincial governments.

PARC is a federal-level organization with a charter to promote, conduct, and coordinate agricultural research at the national level. It operates three research institutes: The Cereal Diseases Research Institute, which conducts research as well as provides research support to other institutions for the development of disease-resistant crop varieties, particularly wheat; and the recently established Arid Zone Research Institute, Quetta, and National Agricultural Research Centre, Islamabad (NARC). NARC is designed to provide facilities for basic research on agriculture and will have a well-maintained agricultural library, a computer-based data centre, a centre for plant introduction and genetic resources, and a centralized facility for instrumentation and repair of laboratory equipment. To strengthen the existing research system, NARC will conduct research in areas of national importance where such research is not currently being done, is seriously inadequate, or can best be done at a central place. PARC also maintains a Vertebrate Pest Control Centre, a Federal Pesticides Laboratory, the National Mycological Herbarium, and the National Insect Museum.

---

[1] Senior Documentation Officer, Pakistan Agricultural Research Council, P.O. Box 1031, Islamabad, Pakistan.

Currently, PARC is operating 17 national level cooperative/coordinated research programs at various provincial and federal institutes, as well as at NARC; coordinating 98 research projects at various research/education establishments; and operating 37 short-term research schemes covering various aspects of agriculture at different institutes.

Research by Subsectors

Research at provincial agricultural research institutions has been largely limited to verification trials and crossbreeding of major crops. Laboratory-oriented research and studies on socioeconomics have been neglected.

Crops

Major research for crops is conducted at four multidisciplinary provincial agricultural research institutes. These institutes comprise commodity sections, coupled with sections on supporting disciplines, including soil and plant chemistry, plant physiology, agronomy, disease and pest control, and food technology. Recently, because of expanding research needs, some of these commodity sections have been raised to the institute level.

In the multidisciplinary category, there are 12 research institutes. The fully operational institutes are the Ayub Agricultural Research Institute, Faisalabad; the Agricultural Research Institute, Sariab, Quetta; the Agricultural Research Institute, Tandojam; the Agricultural Research Institute, Tarnab, Peshawar; the Nuclear Institute for Agriculture and Biology, Faisalabad; and the Atomic Energy Agricultural Research Centre, Tandojam. The remaining six institutes are at different formative stages.

There are 14 monocommodity research institutes: one for wheat; four for cotton; two each for rice and sugar crops; and one each for oilseeds and maize and millets. The Cereal Crops Research Institute, basically the maize and millets research institute, was recently renamed to cover other cereals.

Four institutes cover plant protection: the Cereal Diseases Research Institute; the Commonwealth Institute of Biological Control; the Plant Protection Institute; and the Vertebrate Pest Control Centre. In addition, an insect museum and a mycological herbarium are maintained. A section and a station also exist for research on locust control. The CDRI, VPCC, and other sections are under PARC, whereas PPI is with the Punjab Agriculture Department.

Research on soil and fertilizer use is undertaken by the Rapid Soil Fertility Survey and Soil Testing Institute in Punjab. In other provinces these facilities exist on a smaller scale. The soil and plant chemistry sections at the multidisciplinary institutes also conduct research in this discipline.

Research efforts in agricultural machinery have been seriously lacking and it is only recently that the Agricultural Machinery Institute at Multan was established. The PARC has also established a full division on agriculture machinery at NARC.

Livestock

The development of the livestock sector has received less attention than the crop sector. There are five research institutes, one on livestock production and two each on veterinary sciences and poultry research. The Livestock Production Research Institute at Bahadurnagar has seven divisions: breeding; feeding and management; fodder crops and pastures; economics and marketing; development support communication; health control; and common services. These divisions conduct problem-oriented research.

The Veterinary Research Institute at Lahore has nine divisions: virology; bacteriology; immunology; parasitology; helminthology; biochemistry; poultry vaccines; disease investigation; and epizootiology. The institute also operates a Foot and Mouth Disease Research Centre. The Veterinary Research Institute, Peshawar, established in 1949, has five divisions: microbiology; parasitology; animal production; biological production; and poultry husbandry.

There are two poultry research institutes: one established in 1970 at Karachi and the other established in 1976 at Rawalpindi. Research on pathology, nutrition, breeding and incubation, and economics and marketing is conducted at these institutes.

The provincial livestock development departments maintain about 12 livestock experiment stations in various provinces of the country. The Animal Husbandry Laboratory, Karachi, established in 1939, undertakes disease diagnostic studies and PARC has established a full-fledged livestock division at NARC, Islamabad, to study livestock breeding and the development of feed resources.

Fisheries

The exploitation and development of fish resources has remained low key. A small fisheries institute exists at Qadirabad, under the Directorate of Fisheries, Lahore. The Institute of Marine Biology, the centre of excellence, is located at Karachi University and is under the University Grants Commission.

Forestry

The Pakistan Forest Institute, Peshawar, is the only institute catering to research and education needs in forestry and allied subjects. The Faculty of Agriculture, University of Agriculture, Faisalabad, teaches forestry and range management.

Social Sciences

In addition to a number of university departments and financial institutions, there are four institutions maintaining research and development activities on economics, including agricultural economics and rural sociology. PARC has also recently established a full-fledged Division of Social Sciences at its headquarters to undertake and promote research on various aspects of agricultural economics and rural sociology.

Educational Institutions

There are three agricultural universities, one agricultural college, and one college of veterinary sciences. The Gomal University, D.I. Khan, has a faculty on agriculture. In addition, there are five training institutes. The University of Agriculture, Faisalabad, maintains six faculties: agriculture; agricultural economics and rural sociology; agricultural engineering and technology; animal husbandry; veterinary science; and science. There is one division of extension and three directorates, i.e., Directorate of Research, Directorate of Advanced Studies, and Directorate of Sports. The two colleges, the Barani Agricultural College, Rawalpindi, and the College of Veterinary Sciences, Lahore, are affiliated with Gomal University. The Sind Agriculture University, established in 1976, is operating with three faculties: agriculture; animal husbandry; and agricultural engineering. The faculty of agriculture, University of Peshawar, became the University of Agriculture, Peshawar, in 1981. The universities are under the respective provincial governments. The five training institutes produce field assistants and to some extent make available in-service training facilities to the extension staff.

Lack of Coordination

The institutional structure that has been described involves administrative control by different provincial and federal ministries. This often results in wasteful duplication and makes effective monitoring and coordination difficult. The commodity-oriented approach has resulted in the less efficient use of scarce research manpower and financial resources.

The government, which was concerned with the need to remedy this situation, decided to strengthen the existing system. An important initial step was to strengthen the Pakistan Agricultural Research Council by according it autonomous status and by creating a separate Agricultural Research Division at the federal level.

Resource Allocation System

Finances

Traditionally, funds from the finance department are channeled through the respective ministries to the various institutions. Broadly, funds are of two types, i.e., nondevelopmental (recurring) and developmental (nonrecurring). Developmental funds may provide for capital costs, including buildings and equipment.

The budget in the nondevelopment category is normally a fixed amount made available in bulk on an annual basis. The development budget is linked with short-term research projects and is usually limited to a period of 1–5 years. After achievement of the preliminary objectives, projects of a continuous nature are shifted to the nondevelopment category.

After approval by the institution leader, all projects are critically screened by a provincial research coordination board to assess their suitability and to try to eliminate duplication. Approved projects are then placed before the provincial planning and development department by the concerned ministry for final review and approval and budget allocation. The coordination boards, administrative secretaries, and planning and development departments have progressively increasing sanctioning authority. At the federal level, projects are submitted to the Central Development Working Party. Large projects are approved by the Economic Committee of the National Economic Council (ECNEC), headed by the Minister for Finance, Planning, and Provincial Coordination.

The whole process of project formulation, evaluation, and approval is not very objective and in most cases has resulted in imbalanced growth. Recently, PARC has designed CAREPLANS (Coordinated Agricultural Research Planning System) and CRISP (Current Research Information System of Pakistan) programs that involve modern objective-oriented procedures.

Manpower

Recently, PARC has established a national talent pool and plans to train an adequate number of scientists in the disciplines in which the country is particularly deficient. The services of these scientists will also be loaned to the various institutions on request. PARC is also building manpower at the respective institutions by providing training at local as well as the foreign institutions. Manpower is recruited by advertisement and selection by the Public Service Commissions in the respective provinces.

Total Investment in Agricultural Research

Allocations to agricultural research have steadily increased, but the total investment remains low compared with spending in developed countries. During 1977–78, Rs. 187 million were expended on agricultural research, which represented 0.11% of total GNP and 0.37% of agricultural GDP. Assuming a rate of increase in expenditures on agricultural research of roughly 2.5% per year, the allocation in 1980–81 was estimated as Rs. 200 or 0.07% of GNP, 0.076% GDP, and 0.25% of agricultural GDP.[2]

This drop in the allocation to research as a percentage of agricultural GDP is surprising and discouraging, but can be explained by recent increases in agricultural production that have not been accompanied by corresponding increases in research allocations.

Allocation by Subsectors

Of the total research allocation of Rs. 277.3 million given to educational institutions and research organizations in 1977–78, crops received Rs. 238.4 million (86%), livestock Rs. 29.2 million (11%), fisheries Rs. 6.3 million (2%), and forestry Rs. 3.4 million (1%).

Allocation by Commodities

The trade value of the three major crop commodities at government-fixed procurement rates in 1977–78 was Rs. 15 663 million for wheat, Rs. 6723 million for rice, and Rs. 8937 million for cotton.

The allocation in 1977–78 to the research institutes working on these commodities was Rs. 16.89 million and was distributed as follows: Wheat Research Institute, Faisalabad, Rs. 1.2 million; Rice Research Institute, Dokri, Rs. 4.2 million; Rice Research Institute, Kala Shah Kaku, Rs. 1.75 million; Cotton Research Institute, Multan, Rs. 1.75 million; Cotton Research Institute, Sakrand, Rs. 6.09 million; and Institute of Cotton Research and Technology, Karachi, Rs. 1.9 million. The Cotton Research Institute, Sakrand, was established after

1977–78 and is not considered here. These figures exclude allocations to commodity sections at some of the multidisciplinary research institutes, but these are only a small portion of the total allocations.

Assuming an increase in the budget of 7% from 1977–78 to 1980–81, the total allocation to these three major commodities would be about Rs. 18.1 million or only 0.06% of the trade value of these three commodities.

Allocations to Educational Institutions

During 1977–78, the two agricultural universities and the Faculty of Agriculture, Peshawar University, received Rs. 47 million, or about 17% of the total research and education allocation.

Allocations within Research Institutes

Table 1 presents a percentage breakdown of funds going to different areas within a selected number of agricultural research and education establishments. If salaries and wages are separated from other operating costs, a continual decline in operating costs is seen due to continual increases in scientific staff, their rising salaries, and occasional cuts in contingent allocations to hold down government expenditures.

If we assume that the total amount for contingencies is research money, the operational budget at the nine selected institutes ranged from 20.3% to 30.8%. This percentage includes rents, rates, taxes, and a number of other items that account very roughly for about 50% of the contingencies. Thus, the core allocation for actual research materials is reduced to 10.15–15.4%. The rapidly increasing cost of research materials further aggravates the situation and adversely affects overall research performance. A comprehensive program to correct this situation is required.

Total Agricultural Graduates in Pakistan

The University of Agriculture, Faisalabad; Sind Agriculture University, Tandojam (which also caters to Baluchistan Province); and the Faculty of Agriculture, Peshawar University, produce agricultural graduates. The output of agricultural graduates by these institutions up to 1976–77 was: University of Agriculture, Faisalabad, 7000; Sind Agriculture University, Tandojam, 1967 (1600 for Sind, 367 for Baluchistan); and Agriculture College, Peshawar, 577.

The total number of graduates produced in relation to number of farms is very low. For Pakistan as a whole in 1976–77 there were 9554 graduates serving 4.86 million farms or 48.9 million acres. There is, on average, one graduate for every 514 farms or

---

[2] Data on agricultural resources and agricultural contribution to GNP from Survey of Pakistan 1979–80, Ministry of Finance, Government of Pakistan.

Table 1. Resource allocation by function: percentage of budget allocation to selected agricultural research and educational establishments by primary units of appropriation (1977–78).

	Pay of officers	Pay of establishment	Allowance and honoraria	Contingencies	Total of first 3 columns
Ayub Agricultural Research Institute, Faisalabad	25.50	29.50	14.20	30.80	69.20
Agricultural Research Institute, Tarnab, Peshawar	30.10	37.60	12.00	20.30	79.70
University of Agriculture, Faisalabad	65.55 (staff)		10.79	23.66	76.34
Faculty of Agriculture, University of Peshawar	95.40 (staff)		—	4.60	95.40
Cotton Research Institute, Multan	24.70	12.50	34.80	28.00	72.00
Cotton Research Institute, Sakrand	18.90	25.20	25.50	30.40	69.60
Vegetable Research Institute, Faisalabad	29.31	29.03	19.47	22.19	77.81
Wheat Research Institute, Faisalabad	26.29	26.45	18.66	28.60	71.40
Plant Protection Institute, Faisalabad	27.84	33.77	26.71	11.68	88.32
Sind Horticulture Institute, Mirpurkhas	16.17	54.95	7.48	21.40	78.60
Fisheries Research Institute, Quadirabad, Gujranwala	21.39	31.09	20.66	26.86	73.14

5175 acres.[3] This is about 51 times lower for farms and 4.31 times lower in terms of acreage compared with the United States of America.

Total Scientific Manpower

The total research staff engaged by research and education institutions was 2834 in 1977–78. Of this total, 233 (8.2%) had a Ph.D., 1405 (49.6%) a M.Sc., and 1196 (42.2%) a B.Sc. The universities employed 120 researchers holding a Ph.D., 385 with a M.Sc., and 127 with a B.Sc.; the research institutes 113 at the Ph.D. level, 1020 holding a M.Sc., and 1069 with a B.Sc. The distribution of staff in the different subsectors is shown in Table 2.

Table 2. Distribution of research staff in the different agricultural sectors (1977–78).

	Ph.D.	M.Sc.	B.Sc.
Crops	184	1208	930
Animal husbandry[a]	40	138	210
Forestry[a]	6	40	36
Fisheries[a]	3	19	20

[a]Includes staff working in the universities.

[3] Data on agricultural graduates and number of farms and farm areas from Dr M. Sami Khan Abid, Agricultural Graduates: Importance and Utilization, The Varsities, March–April 1978.

Classification of Manpower by Commodity

In the absence of detailed information, the classification of manpower by commodity is based on the number of researchers working at different commodity institutes (Table 3).

PARC Program on Resource Allocation

To provide a basis for national coordinated agricultural research planning and effective management of research resources, the Pakistan Agricultural Research Council has instituted an automated system for storing and retrieving information on research projects in Pakistan.

The system known as the Current Research Information System of Pakistan (CRISP) uses a computer to store, for each research project, descriptive information such as title, principal investigator, performing organization, objectives, methodology, progress reports, and expenditures. In addition, it describes each project in terms of research problem area, field of science, commodity, and research activity. It will thus be possible to retrieve from the system summaries of research activities by commodity, field of science, activity, research problem area, location, subjects, etc.

Inputs into the system are made on CRISP I, II, and III forms covering different items including information on manpower and financial allocations as well as project objectives, performers, and location of research activity and in addition the methodology and the progress report. The output can take many forms including reports and summaries. The outputs that are expected immediately are: (1) a

	Ph.D.	M.Sc.	B.Sc.	Total
Cotton				
Cotton Research Institute, Multan	4	19	15	38
Cotton Research Institute, Sakrand	1	17	6	24
Institute of Cotton Research and Technology, Karachi	1	10	24	35
Rice				
Rice Research Institute, Kala Shah Kaku	4	21	11	36
Rice Research Institute, Dokri	1	23	10	34
Wheat				
Wheat Research Institute, Faisalabad	2	31	3	36

[a]Excludes staff in commodity sections at multidisciplinary institutes.

summary of resources expended on major commodities, location, and research problem areas; (2) a summary of progress on each major commodity, location, and research problem areas; and (3) a "directory of current research in agriculture," a summary of resources expended in all agricultural institutes in Pakistan.

The computer program for CRISP has been prepared and pilot output has been obtained using input from 150 projects. The implementation of CRISP on a national level is in progress.

Uncertainties

The data on manpower and budget allocations presented in this paper are based on information collected in 1977–78, under a collaborative effort by the Pakistan Agricultural Research Council and the National Science Council of Pakistan. These data have a few uncertainties. The budget information on the allocations by commodity and sector and the data on budget and staff at the multidisciplinary institutions and at certain organizations could not be separated in the absence of detailed information. Similarly, the allocations could not be separated for certain institutions undertaking both research and development activity. A more elaborate study is required to better understand the research system in Pakistan.

Resource Allocation to Agricultural Research in Sri Lanka

Y.D.A. Senanayake and H.M.G. Herath[1]

The agricultural sector occupies a dominant position in the economy of Sri Lanka. Nearly 80% of the land area is used for agricultural production and forestry. Approximately 48% of the land is under annual, semiperennial, and perennial crops and the potential exists to increase production as more land is developed and its infrastructure established. The agricultural base has been developed through crop production; whereas the livestock and fisheries subsectors are comparatively smaller than the crop subsector. The food balance sheet of Sri Lanka shows that livestock and fisheries together contribute only 6–7% of the protein intake of the people; whereas crop products contribute the remainder.

The population of Sri Lanka was 14.8 million in mid-1980 and the average rate of growth during the 6-year period 1975–1980 was 1.73%. The latest trends indicate that nearly 51% of the population is directly employed in agriculture. If the agricultural manufacturing, processing, trade, and servicing subsectors of the economy are also considered as direct and indirect employment in agriculture this increases to about 70% of the population.

For many years, 75–80% of the export income was derived from the three main export crops tea, rubber, and coconuts. This proportion declined to 63% in 1980 due to fluctuations in production and market prices. Nevertheless, these three crops continue to be the main foreign exchange earners for the country.

Recently, the relatively rapid growth in population has increased the demand for food and has also exacerbated the unemployment problem. The goal of self-sufficiency in food has not yet been achieved. In addition, the varying fortunes of the export sector have exerted a deleterious influence on the economy. These problems require urgent attention. Their association with the agricultural sector suggests that the solution in the short run needs to be evolved in the agricultural sector itself. One of the main ways by which the severity of the problems could be mitigated is to enhance the productivity of the agricultural sector. An increase in the productivity of the agricultural sector could be achieved through a strong research organization.

The Agricultural Research Sector

Sri Lanka has attempted to speed up the pace of development, through technological change in the agricultural sector, by investment in agricultural research. Agricultural research in Sri Lanka had its origins in the Department of Agriculture at Peradeniya. Soon after the turn of the century when the plantation sector consisting predominantly of tea, rubber, and coconuts expanded, three commodity research institutes were established to undertake research on these three crops. Research on annual, semiperennial, and perennial crops of small farmers as well as livestock continued to be the responsibility of the Department of Agriculture. Research on all crops functioned under the Ministry of Agriculture.

For 25 years after independence there was no change in this functional separation. But during the seventies the responsibility for research in the agricultural sector was handed over to several ministries. Today as many as 11 Ministries and 24 units under them are responsible for the development of research on commodities or fields directly related to them (Table 1). In addition, specialized units or bodies in some of these Ministries such as the National Science Council and the Atomic Energy Authority of the Ministry of Industries and Scientific Affairs lend support to agricultural research. But there is no central Council or Authority to coordinate the research activities of different Ministries and to formulate and implement policy guidelines. Its absence could soon lead to wasteful allocations of

---

[1] Dean of Agriculture and Lecturer in Agricultural Economics, respectively, Faculty of Agriculture, University of Peradeniya, Sri Lanka.

Table 1. Ministries and their agencies responsible for research in the agricultural sector.

Ministry of Agricultural Development and Research
    Department of Agriculture — (Research Division)
    Department of Minor Export Crops — (Research Division)
    Agrarian Research and Training Institute
    Sugar Corporation — (Sugarcane Research Institute)
    Rice Processing and Development Centre
    Agricultural Diversification Unit
Ministry of Plantation Industries
    Tea Research Institute
    Rubber Research Institute
    Cashew Corporation — (Research Division)
    Silk and Allied Products — (Research Division)
Ministry of Coconut Industries
    Coconut Research Institute
Ministry of Rural Industrial Development
    Department of Animal Production and Health — (Research Division)
Ministry of Lands and Land Development
    Forest Department — (Research Division)
    Irrigation Department — (Research Division)
Ministry of Fisheries
    Department of Fisheries — (Research Division)
Ministry of Higher Education
    Faculty of Agriculture
    Postgraduate Institute of Agriculture
Ministry of Mahaweli Development
    Mahaweli Research Unit
Ministry of Industries and Scientific Affairs
    Ceylon Institute of Scientific and Industrial Research
    National Science Council
    Atomic Energy Authority
Ministry of Finance and Planning
    Central Bank — (Economic Research Division)
Ministry of Plan Implementation
    Department of Census and Statistics

resources due to the competing claims of different Ministries for resources that are becoming scarce and expensive. In spite of the changes brought about in the seventies, the Ministry of Agriculture still has the largest research arm with multicrop/multidisciplinary research centres and smaller research stations located in different agro-ecological regions across the country. The commodity research institutes for tea, rubber, coconut, sugarcane, and minor export crops have one principal research station each and 1–3 substations. Sericulture and cashew, being relatively new crops with embryonic research divisions, have begun some research in one of their principal plantations. The research arm of the Department of Animal Production and Health has as its principal station the Veterinary Research Institute at Peradeniya and other subcentres distributed on animal production farms of the department.

Investment of Funds on Agricultural Research

The treatment of agricultural research and extension in an economic framework is of relatively recent origin. Thus, a systematic compilation of research expenditure data for individual countries is not widely available. Data on agricultural research expenditures have been compiled to obtain a better perspective of agricultural research investment in Sri Lanka.

Data were compiled from many sources: from the annual budget estimates of most institutes; from annual reports; and from direct communication with certain research institutes. In most cases, budget data refer to actual expenditures, but for 1980 only estimates are available.

The allocation of funds is investigated in only the most important of the institutes: the Department of Agriculture; the Department of Animal Production and Health; the Agrarian Research and Training Institute (ARTI); the Faculty of Agriculture; the Postgraduate Institute of Agriculture; the Tea, Rubber, and Coconut Research Institutes; the Sugarcane and Cashew Corporations; and the Department of Minor Export Crops.

Research in the Department of Agriculture straddles several crops such as rice, pulses, condiments, fruits, and vegetables and several disciplines such as agronomy, breeding, physiology, and chemistry. In general, a crop-wise or a discipline-wise classification of research expenditure is not maintained in the Department of Agriculture. The research expenditure incurred is subsumed under items such as salaries, travel, and communications. The data available thus comprise the total expenditure for all crops and all disciplines. An attempt is being made to obtain an approximate picture of allocation of funds by crops in the Department of Agriculture. Hopefully some results will be available in due course. The ARTI undertakes mostly socioeconomic research. The research expenditure in the Faculty of Agriculture and the Postgraduate Institute of Agriculture forms an insignificant proportion in the overall allocation and a finer breakdown of this into crops was felt to be unwarranted.

Crop-wise research expenditure was readily available for tea, rubber, coconut, sugarcane, and cashew. For tea, rubber, and coconut, the establishment of separate institutes designed specifically for their research facilitated the recording of research expenditure data on a crop-wise and discipline basis. For sugarcane and cashew research, expenditure data are available from the Sugar Corporation and the Cashew Corporation although research in these crops is still in its infancy.

Trends in Investment in Agricultural Research

Table 2 presents some figures for annual expenditures on research for the years 1975–80. The total research expenditure under all categories is also presented. Research expenditure on agriculture increased steadily from Rs.27.63 million in 1975 to Rs.91.02 million in 1980. The increase in 1980 represents nearly a 230% increase over the expenditure in 1975.

The distribution of the total research expenditure among the crops or/and institutions indicates that the Department of Agriculture constitutes the largest single component in total agricultural research in Sri Lanka. The research expenditure in the Department of Agriculture was Rs.9.05 million in 1975. This rose to 9.62 million in 1976, and then increased notably in 1977 (by approximately 145%). This increasing trend continued until 1980.

Expenditure on animal production research was reported to be Rs.4.14 million in 1975. In 1976 it increased to Rs.7.18 million, but it declined both in 1977 and 1978. In 1979 the expenditure increased to 7.12 million and in 1980 a record increase of approximately 100% over that in 1979 was recorded. The total research expenditure of the Agrarian Research and Training Institute (ARTI) hovered between Rs.2.52 million and Rs.4.48 million during the 1975–80 period.

Total research expenditure on tea was 1.3 million in 1975. This has steadily increased during 1975–80 (excepting 1979) to Rs.6.8 million. The decline in 1979 was caused by a change of the management of the estates. A similar pattern of research expenditure was observed for rubber. A steady increase in the research expenditure for coconuts is indicated. Research on sugarcane and cashew is of recent origin. The expenditure on sugarcane has remained stationary at approximately Rs.3 million annually. Cashew research represents an insignificant component in overall expenditure. For minor export crops, research expenditure has steadily increased, with a substantial increase occurring in 1980 of approximately 81% over the 1979 level.

To further explore the allocation of research funds for agriculture, the shares of gross national product (GNP) and agricultural gross national product (AGNP) allocated to agricultural research were computed (Table 3). The research expenditure as a percentage of GNP was 0.107 in 1975. This percentage increased slightly to 0.149 in 1978 and more or less flattened out for the rest of the period. The ratio of research expenditure to AGNP was 0.363 in 1975 and rose to 0.647 in 1979.

Most of the institutes allocate funds for other activities such as extension and administration. The

Table 2. Agricultural research expenditure (in millions of rupees) in Sri Lanka.[a]

Commodity/Institute	1975	1976	1977	1978	1979	1980
Tea[b]	1.26	1.67	2.62	5.89	3.63	6.86
Rubber[c]	2.94	3.97	5.04	7.46	5.74	8.74
Coconut[d]	2.67	2.85	3.05	4.05	5.96	7.09
Sugar[e]	2.25	2.81	2.61	2.97	2.94	2.93
Cashew[f]	Nil	Nil	0.04	0.08	0.11	0.14
Minor export crops[g]	2.79	3.22	2.28	2.90	5.34	9.68
Animal production[h]	4.14	7.18	6.97	5.93	7.12	14.64
Department of Agriculture[i]	9.05	9.62	23.61	28.31	36.18	35.31
Faculty of Agriculture and PGIA[j]	—	0.27	0.31	0.33	0.41	1.15
Agrarian Research and Training Institute[k]	2.52	2.88	1.61	2.03	3.16	4.48
Total	27.63	34.47	47.96	59.95	70.59	91.02

[a] Total of recurrent and capital expenditure.
[b] Source: Project budgets of the Tea Research Institute of Sri Lanka, Talawakelle, 1975–80.
[c] Source: Program expenditure reports of Rubber Research Institute of Sri Lanka, Agalawatte, 1975–80.
[d] Source: Annual draft estimates of the Coconut Research Institute of Sri Lanka, Lunuwila, 1975–80.
[e] Source: Sri Lanka Sugar Corporation, Colombo.
[f] Source: Sri Lanka Cashew Corporation.
[g] Source: Estimates of the revenue and expenditure of the Government of Sri Lanka, 1975–80 (crops include coffee, cocoa, pepper, etc.).
[h] Source: Estimate of the revenue and expenditure of the Government of Sri Lanka, 1975–80.
[i] Source: Office of the Deputy Director of Agriculture (Research), Department of Agriculture, Peradeniya.
[j] Source: Dean, Faculty of Agriculture, University of Peradeniya.
[k] Source: Agrarian Research and Training Institute, Colombo.

Department of Agriculture, for example, has five major divisions in its organization: research; extension; education and training; administration; and planting materials production (farms division). The allocation of funds among these divisions is examined in Table 4. Among the five divisions, seed and planting materials production has received the highest share, approximately 38%, except in 1977 when it declined to 26%. Agricultural extension received the next largest share in 1975 and 1976. The relative proportion for research was lower than for extension in 1975 and 1976 but this pattern reversed in 1977. The percentage allocations for administration and education and training are relatively low.

The allocation of funds among the three divisions of research, extension, and administration in animal production are also given in Table 4. Here the data are complete only after 1977. The proportion spent for extension is high and has varied between 43.8 and 52.1% during the study period. Research comes next with an allocation of approximately 30% of the total expenditure in the department. Administrative expenditure is relatively low. For minor export crops the expenditure on research in relation to extension is high. The administrative expenditure is relatively low.

In tea, relatively little is spent on extension activities. Most of the expenditure is on research and administration and the relative allocation for administration is very high although it appears to have declined after 1977. For rubber, the relative importance of the various divisions has fluctuated. However, resarch in rubber has received at least 30% of the total allocation and in 1979 and 1980 the percentage for research reached approximately 50%. The expenditure for research in coconuts is approximately 30% of the total allocation and ex-penditures for extension activities and administration expenditures are also high.

Sources of Funds for Agricultural Research

Agricultural research is generally supported by several organizations in addition to the national government. The sources of finance for agricultural research in Sri Lanka are summarized in Table 5. An examination of the sources of finance for research conducted by the Department of Agriculture indicates that in 1975 the government allocation (consolidated fund) provided 91.2% of the support and 8.8% was from foreign sources (mainly West Germany). This pattern was similar in 1976. In 1977, the foreign aid component rose to Rs.10.2 million. Although the expenditure by the Sri Lankan government also increased in 1977, foreign aid contributed 43.1% of total research expenditure. Two main sources of foreign funding were forthcoming in 1977. These were Rs.4.76 million from IDA/IBRD and Rs.3.284 million from the UNDP/Soyabean project. In 1978 and 1979 the relative proportion of government expenditure to foreign funds was maintained. In 1980, there was a slight decline in funds from foreign sources. For animal production research, foreign aid has always been the dominant financial source. This support has fluctuated between 74.4% in 1974 and 87.8% in 1980. The major sources of foreign funding are CIDA and IDA.

For tea, the cess is the main source of finance. Between 85 and 90% of the funds used for tea research are generated from the cess and the additional expenditures are covered by commercial activities. Tea research is thus generally self-financing. For rubber, the cess is the dominant component although it is not as high as for tea. This has

Table 3. Total agricultural research expenditure (in millions of rupees) in Sri Lanka as a percentage of gross national product (current prices), gross domestic product (current prices), and agricultural gross product (current prices).

	1975	1976	1977	1978	1979	1980
Total expenditure for research (TE)	27.63	34.47	47.96	59.95	70.59	91.02
Gross national product[a]	25746	28216	34681	40098	48885	61807
Gross domestic product[a]	25959	28494	34933	40335	49125	62246
Agriculture gross national product[a]	7617	7983	10193	11355	10902	14210
TE as % of GNP	0.107	0.122	0.138	0.149	0.144	0.147
TE as % of GDP	0.106	0.121	0.137	0.148	0.143	0.146
TE as % of AGNP	0.363	0.432	0.471	0.528	0.647	0.641

[a] Source: Review of Economy, Central Bank of Ceylon, 1975–80.

Table 4. Expenditure on research, extension, and administration by commodity (in millions of rupees).

	1975		1976		1977		1978		1979		1980[a]	
	Amount	%	Amount	%	Amount	%	Amount	%	Amount	%	Amount	%
Department of Agriculture[b]												
Research	9.05	14.8	9.62	11.9	23.61	23.7	28.31	20.5	36.18	23.2	35.31	17.7
Extension	16.19	26.6	24.15	30.1	20.58	20.6	22.79	16.5	27.71	17.8	38.09	19.1
Seed and planting material production	23.30	38.2	30.36	37.8	25.96	26.0	53.81	39.1	58.79	37.7	75.59	37.8
Education and training	6.47	10.6	6.25	7.8	10.79	10.8	16.98	12.3	13.21	8.5	27.29	13.7
Administration and others[c]	5.98	9.8	9.98	12.4	18.79	18.8	15.63	11.4	19.91	12.8	23.17	11.6
Total	60.99	100	80.36	100	99.73	100	137.52	100	155.81	100	199.45	100
Animal Production												
Research	4.14	100	7.18	100	6.97	34.1	5.93	31.3	7.12	28.3	14.64	36.8
Extension and advisory	—	—	—	—	8.96	43.8	8.58	45.3	13.31	52.6	13.13	33.1
Administration	—	—	—	—	4.5	22.1	4.44	23.4	4.84	19.2	11.98	30.1
Total	4.14	100	7.18	100	20.43	100	18.95	100	25.27	100	39.75	100
Minor Export Crops												
Research	2.79	71.8	3.22	72.6	2.28	64.6	2.91	65.4	5.37	72.9	9.67	75.3
Extension	0.51	12.9	0.70	15.8	0.72	20.3	0.96	21.7	1.33	18.1	1.31	17.9
Administration	0.59	15.3	0.52	11.8	0.53	15.1	0.57	12.8	0.68	9.3	0.86	6.7
Total	3.89	100	4.43	100	3.52	100	4.43	100	7.34	100	12.84	100
Tea												
Research	1.26	19.0	1.67	22.9	2.62	39.9	5.89	50.4	3.63	25.3	6.86	37.7
Extension and advisory	0.26	3.9	0.36	4.8	0.36	5.4	0.54	4.6	2.38	16.6	3.96	21.8
Administration	5.11	77.1	5.21	72.2	3.59	54.7	5.24	44.8	8.34	58.2	7.38	40.6
Total	6.63	100	7.24	100	6.57	100	11.67	100	14.53	100	18.19	100
Rubber												
Research	2.94	50.0	3.97	44.7	5.04	22.7	7.47	29.5	5.74	52.5	8.74	49.6
Extension and advisory	—	—	1.69	20.1	1.84	8.2	1.95	7.7	2.31	21.1	2.85	16.2
Administration	2.17	36.8	2.39	27.1	2.73	12.3	2.86	11.3	0.65	5.9	5.09	28.9
Others[d]	0.78	13.3	0.63	7.4	12.57	56.7	13.14	51.8	2.23	20.2	0.91	5.2
Total	5.88	100	8.87	100	22.18	100	25.39	100	10.93	100	17.59	100
Coconut												
Research	2.67	28.2	2.85	27.8	3.05	27.8	4.05	24.8	5.96	32.1	7.09	27.8
Extension and advisory	2.89	30.5	2.78	26.5	3.33	30.4	6.69	41.1	1.98	10.6	9.21	36.1
Administration	3.92	41.3	4.85	46.3	4.57	41.7	5.81	35.6	10.61	57.0	8.89	35.0
Total	9.47	100	10.47	100	10.95	100	16.29	100	18.62	100	25.48	100

[a] Estimated values.
[b] All crops studied by the Department of Agriculture.
[c] Includes finance, pilot projects, and engineering division.
[d] Includes statistical services and library and publication services, group processing centres, and plantation division.

Table 5. Summary of sources of finance (in millions of rupees) for research in Department of Agriculture, Department of Animal Production and Health, Tea Research Institute, Rubber Research Institute, and Coconut Research Institute.

Source of finance	1975		1976		1977		1978		1979		1980	
	Amount	%	Amount	%	Amount	%	Amount	%	Amount	%	Amount	%
Department of Agriculture[a]												
Consolidated fund	8.251	91.2	9.161	95.3	13.452	56.9	15.861	56.1	18.791	51.9	21.412	60.6
Foreign aid	0.795	8.8	0.464	4.7	10.166	43.1	12.449	43.9	17.398	48.1	13.894	39.4
Dept. Animal Prod. and Health[b]												
Consolidated fund	1.059	25.6	1.575	21.9	1.08	15.5	1.218	20.5	1.291	18.1	1.789	12.2
Foreign aid	3.081	74.4	5.601	78.1	5.892	84.5	4.716	79.5	5.827	81.9	12.847	87.8
Tea Research Institute[c]												
Cess + Government	7.52	89.5	7.59	75.7	1.65	20.7	10.0	85.8	13.55	94.5	16.11	88.5
Commercial activities	0.87	10.1	2.19	21.7	6.19	77.9	1.51	12.9	0.54	3.8	1.37	7.6
Others[d]	—	—	0.26	2.6	0.09	1.2	0.16	1.3	0.23	1.7	0.71	3.9
Rubber Research Institute[e]												
Cess	3.30	56.1	5.60	62.6	6.00	25.6	4.80	19.9	7.20	56.5	12.00	63.4
Government	0.12	2.2	1.88	21.1	1.99	8.5	3.04	15.2	4.77	37.5	1.94	10.3
Sale of products	1.28	21.8	0.90	10.1	13.74	58.6	12.24	61.2	—	—	3.24	18.1
Others[f]	1.17	19.9	0.56	6.3	0.71	3.1	0.73	3.6	0.77	6.1	1.57	8.3
Coconut Research Institute[g]												
Cess	0.50	5.3	7.45	71.1	7.73	73.8	1.52	9.3	—	—	—	—
Government	7.09	75.3	0.50	4.8	0.50	4.6	10.75	65.9	13.78	74.1	20.69	81.2
Sale of products	0.91	9.5	1.58	15.1	1.77	16.2	2.97	18.3	3.61	19.3	3.29	12.9
Others[h]	0.98	10.3	0.94	8.9	0.95	8.6	1.04	6.4	1.23	6.6	1.52	6.9

[a] Source: Office of the Deputy Director of Agriculture (Research), Department of Agriculture, Peradeniya.
[b] Sources: Annual estimates of Department of Agriculture, Sri Lanka, 1975–80; Annual estimates of Department of Animal Production and Health, 1977–80.
[c] Source: Project budgets of TRI of Sri Lanka, Talawakelle, 1975–80.
[d] Includes interests, royalties, and rentals.
[e] Source: Program expenditure reports of RRI of Sri Lanka, 1975–80.
[f] Includes interest on investments and institute loans, repayment of loans, and sundry receipts.
[g] Source: Annual draft estimates of CRI of Sri Lanka, 1975–80.
[h] Includes administration and motor vehicle working account.

varied somewhat as there was a sharp decline in 1977 and 1978. The sale of products has generated sufficient funds and this in fact compensated for the drop in the cess in 1977 and 1978. Government contribution is the dominant component in coconut research. The cess appears to have provided sufficient funds in 1976 and 1977.

Concluding Remarks

Several aspects relating to the allocation of financial resources in agricultural research in Sri Lanka have been discussed. These results form part of a broader study being undertaken on resource allocation to agricultural research in Sri Lanka. The allocation of research personnel is also being investigated although the results are not yet available. A logical extension to this study is to explore the productivity of agricultural research in Sri Lanka. It is hoped that the present study provides a clearer picture of resource allocation in research and will enable policymakers to identify deficiencies in the system. This would allow the research structure to increase its contribution to higher productivity in the agricultural sector and thereby improve the well-being of farmers.

The authors acknowledge the research grant from SEARCA, Philippines, that supported this study. We also acknowledge the cooperation extended to us by P.C. Munasinghe of IDRC, who originally suggested this study, and thank all the directors of departments, research institutions, and corporations who responded speedily to our requests for information.

Research Priorities and Resource Allocation in Agriculture: The Case of Colombia

Fernando Chaparro, Gabriel Montes, Ricardo Torres, Alvaro Balcázar, and Hernán Jaramillo[1]

The purpose of this paper is to analyze the present experience of formulating a National Plan for Agricultural Research in Colombia. Emphasis is placed not on the substantive content of the plan (i.e., objectives, strategy, and proposed research programs) but on the methodological aspects involved in its formulation. Special attention is given to the criteria and methodological framework that are being used in the process of identifying technological requirements and research priorities (both in terms of agricultural products and research topics or issues) as instruments of resource allocation in this sector.

The first section of the paper provides general information on the present situation and orientation of agricultural research activities in Colombia. The objective is to give a very broad characterization of the present research effort within the country in terms of the areas it covers and the financial and human resources dedicated to it.

The second section analyzes the general methodological framework for the identification of research priorities that is presently being used in the formulation of the National Plan for Agricultural Research in Colombia. The approach that is being used is characterized by two phases: (1) identification of socioeconomic priorities in terms of products or problem areas and (2) determination of technological requirements and research needs for selected products or problem areas.

The Colombian experience with respect to the implementation of these two phases is analyzed in the last two sections of the paper. The institution that has been responsible for the formulation of this research plan in Colombia has been the Instituto Colombiano Agropecuario (ICA), with active collaboration from COLCIENCIAS (Colombian Fund for Scientific and Technological Research) and the National Planning Agency (D.N.P.). The strategy and methodology used in the formulation of this plan was developed by the research people of ICA.

Agricultural Research in Colombia: Institutional Infrastructure and Present Orientation

This part of the paper presents the results of a study conducted by the International Development Research Centre (IDRC) on the way in which resources (financial resources in particular) are allocated for agricultural research in Colombia.[2] The study focused on six institutions and the university sector. The institutions analyzed were: Instituto Colombiano Agropecuario (ICA) (The Colombian Agricultural Institute); Centro Nacional de Investigaciones del Café (CENICAFE) (The National Coffee Research Institute); Corporación Nacional de Investigación y Fomento Forestal (CONIF) (The National Research and Forestry Development Corporation); Corporación Autónoma Regional del Cauca (CVC) (The Cauca Valley Corporation); Centro Internacional de Agricultura Tropical (CIAT) (The International Tropical Agriculture Centre); and Instituto Nacional de los Recursos Renovables y del Ambiente (INDERENA) (The National Institute of

---

[1] Regional Director, Centro Internacional de Investigaciones para el Desarrollo, Bogota, Colombia; Chief, Departamento Nacional de Planeacion, Bogota, Colombia; Agricultural Program Coordinator, COLCIENCIAS, Bogota, Colombia; Researcher, Unidad de Estudios Especiales, Banco Ganadero, Bogota, Colombia; and Programing Assistant, Centro Internacional de Investigaciones para el Desarrollo, Bogota, Colombia, respectively.

[2] IDRC. 1980. Project ARIAL. Asignación de recursos para investigación en América Latina. Colombia: estudio de caso.

National Resources and the Environment). Information on universities doing some type of agriculturally related research was also examined and summarized. ICA was the most important institution studied; being the leading agricultural research centre in the country. It is important to note that this study only considered financial resources spent on agrobiological research.

Expenditure in Agricultural Research

Research and Development Expenditure at the National Level

Table 1 shows the total amount of financial resources that the six institutions studied spent on research from 1972–1976.[3] ICA's share of total research expenditure during this period was 83.5%. However, the table indicates that ICA's share has declined in recent years; in 1973, it accounted for 84.8% of total resources spent on agricultural research, but in 1976 this percentage dropped to 80.3%. During this same period, CENICAFE occupied second place in terms of research expenditure with 10.0%. INDERENA spent an average of 3.0% of total resource expenditure during this period and universities accounted for 3.6%. The CVC share made up no more than 0.9% of the total. Table 1 also shows that although total agricultural research expenditure increased from 1972–1976 (in current values), in real terms (at constant 1970 values) there has been an overall decline in the amount of funds allocated for research.[4]

Table 2 shows the breakdown of agricultural research expenditure in terms of crops and agricultural products, as well as the relationship between research expenditure and value of production for each product. In most cases, the percentage of research expenditure over the value of production is less than 0.20%, with a few extreme exceptions (i.e., oats and sheep) in which the high percentage is due to the small value of that crop's production in the country. In those cases, even a modest research expenditure represents a high percentage in terms of this relationship.

Two additional factors should be pointed out with respect to Table 2. Firstly, the research expenditure figures for the different crops slightly underestimate the investment level in each crop because these amounts only include the cost of the respective research programs but do not include the maintenance costs and investments related to the research stations and centres in which the programs are carried out. This latter aspect appears as a separate expenditure in Table 2. At the aggregate level, total agricultural research expenditure represents 0.33% of the total value of agricultural production in Colombia (with only slight variations between 1972 and 1976).

Secondly, a more significant relationship to analyze is that of agricultural research expenditure as a percentage of the agricultural gross domestic product (GDP) because the latter only includes the value added by this sector. Nevertheless, the breakdown of agricultural GDP in terms of the different crops and agricultural products is not available.

At the sectoral level, Table 3 shows the evolution of the relationship between total agricultural research expenditure and the GDP of the country. This table clearly shows the deterioration of the proportion of agricultural GDP that is allocated to research in this sector. In 1972, this proportion was 0.32%, which was substantially higher than the overall relationship between total national research and development expenditure (for all sectors) and total GDP (estimated by COLCIENCIAS to be 0.20% in 1972). By 1976, this situation had changed drastically, with agricultural research expenditure dropping to 0.22% of agricultural GDP. A somewhat less negative evolution is observed with respect to total GDP (Table 3) and total value of agricultural production (Table 2).

Distribution of Research and Development Expenditure in ICA

Table 4 shows how the distribution of ICA research funds has evolved from 1970–1978. Research activity has tended to decline. Even though total ICA expenditures have increased in real terms, allocations for research have dropped in real terms by $21 000 000 or 17.0%. Research went from constituting 41.1% of the total ICA budget in 1970 to 27.7% in 1978.

A breakdown of the total ICA budget during the period in question shows that this institution has been increasingly assigned more duties but has not received a proportionate increase in budget funds. Consequently, the institute's departments compete for available resources; research, formerly the most important ICA activity, has been negatively affected by this situation in terms of being able to sustain the pace of research projects, undertaking new projects in response to emerging agricultural needs, and losing qualified staff.

---

[3] To convert from Colombian pesos to U.S. dollars, the following rates of exchange should be used for the different years: 1970, $18.45 Colombian pesos for U.S. $1 (this rate should be used for all amounts given in constant 1970 values); 1972, $21.87; 1974, $26.06; 1976, $34.70; and 1978, $39.10.

[4] This does not include CIAT expenditures in this area because CIAT is an international agency and the information would distort the national research picture.

Table 1. Total expenditure on agricultural research (thousands of Colombian pesos).

Institution	1972	1973	1974	1975	1976
ICA	151200	175500	188100	236700	266700
CENICAFE	—	15674	23881	31584	37227
INDERENA	—	9047	9481	9503	9023
CONIF	—	—	—	3053	2813
CVC	—	—	—	1928	3136
Universities	4576	6776	7143	10812	13401
Total	155776	206997	228605	293580	332300
Total (in constant 1970 values)	124422	135469	117233	124610	114114

Source: IDRC. 1980. Project ARIAL. Asignación de recursos para investigación en América Latina. Colombia: estudio de caso.

ICA research can be divided into two categories: agricultural and livestock. These categories can be further divided into basic research and research on specific products. Basic research, which will not be discussed here, includes crop production, grasses and fodder, and special projects.

Table 5 shows that agricultural research represented more than half of the total resources spent by ICA on research. Product research, rather than basic research, predominates in both the agricultural and livestock categories. A brief discussion of these research areas follows.

(1) Agricultural product research.[5] The most important subgroup, in budget terms, in the agricultural product research category is grains and cereals. Table 6 indicates that the maize and sorghum program is the main program[6] because its share in total ICA budget expenditure for the given period is the highest. Rice and wheat are second and third, respectively, after maize and sorghum.[7] These are the most important products in economic terms when you consider the area sown with them and their production value. These products also receive the highest research priority.

The potato and cassava program has also received significant budget allocations, placing it second after the cereal and grain program. These two products also have a substantial share of production value. Over the last 5 years, ICA has increased budget allocations for the fruit and vegetable program because it covers essential food items regarded as high priority in integral rural development plans and food and nutrition programs.

Research on ''panela'' (sugar loaf) also appears important among total research expenditures as a result of the concern the government has shown for this basic subsistence crop that is grown in five regions of the country.

Finally, it is important to note that although some products, such as bananas, represent a considerable part of the production value, ICA has not given them top research priority. This particular commercial crop (bananas) is primarily used for export.

(2) Livestock research by product. The dairy and beef programs account for a significant share of ICA research funds spent on livestock programs/products (Table 7). The pork program is third in terms of budget allocations for livestock research but shows the highest growth rate, whereas the products that are first and second show negative growth rates.

(3) Basic agricultural and livestock research. Tables 8 and 9 provide information on basic research in these two fields. The soil and plant pathology programs are first in basic research. Entomology and plant physiology are allotted a smaller share of funds for basic research. Generally speaking, priority has been given to those disciplines that aim at controlling both plant and animal pests and diseases.

Implicit Research Priorities for Agricultural Products in ICA

On the basis of Table 6, implicit research priorities for agricultural products can be identified according to the amount of funds spent: (1) high priority: maize and sorghum, perennial oleaginous products, potatoes and cassava, fruits and vegetables, and rice; (2) medium priority: legumes and

---

[5] This analysis of research expenditures and economic importance does not include coffee, which is the principal agricultural product in the economy. The National Federation of Coffee Growers conducts research on this product, which receives the largest amount of research funds.

[6] Estimates indicate that almost 80% of the activities in this program are focused on maize.

[7] Although wheat is an important cereal, it is not very important in terms of the amount of funds allocated to it for research. At the economic level, its contribution to production value is not significant. Maize has fundamentally become an imported product.

Table 2. Relationship between research expenditures and value of production by agricultural product (thousands of Colombian pesos).

Product	Value of production (A)					Research expenditure (B)					B/A (%)				
	1972	1973	1974	1975	1976	1972	1973	1974	1975	1976	1972	1973	1974	1975	1976
Coffee	6701590	8540240	10446400	13707100	27189640	350[a]	10155	12294	16123	15506	—	0.119	0.118	0.118	0.057
Rice	1880230	3808710	5668660	6315580	6405360	2834	2979	2884	4380	5452	0.151	0.078	0.051	0.069	0.085
Oats	—	1760	2160	3900	3700	758	1667	1147	637	704	—	94.716	53.102	16.333	19.027
Barley	201586	247476	354147	660386	444886	758	1021	897	925	704	0.376	0.413	0.253	0.140	0.158
Maize	1749450	2460130	2662600	2964820	4288590	4934	5602	6901	6732	7522	0.226	0.173	0.185	0.161	0.125
Sorghum	432390	778400	1069970	1205660	1750000										
Wheat	173968	202358	264364	251527	290554	1501	1931	2790	3648	4021	0.863	0.954	1.055	1.450	1.384
Potato	1190880	5703490	2241580	5335440	4478260	4053	4196	4292	5994	6724	0.098	0.050	0.063	0.050	0.064
Cassava	2945730	2635360	4579320	6572290	6045710										
Yam	—	182597	178448	243975	281766	—	—	—	—	—	—	—	—	—	—
Sugarcane	920112	—	—	—	—	1137	1268	1071	1675	270	—	—	—	—	—
"Panela"	1987290	2814240	2524120	2710040	7562820	—	—	—	1772	4749	—	—	—	0.065	0.063
Cotton	2107470	2948910	3937790	4120030	6894300	2436	2258	2069	2625	3011	0.116	0.077	0.053	0.064	0.044
Sesame	147698	110555	177315	239685	271411	4058	5117	5158	7417	7784	0.839	0.532	0.319	0.364	0.494
Peanuts	—	3213	5313	14250	18260										
African Palm	—	427328	740765	612304	683038										
Soybean	336210	421562	691638	1172180	603900										
Vegetables and fruits	—	—	—	—	—	4456	5005	4578	6878	7754	—	—	—	—	—
Sisal Hemp	—	—	—	—	—	—	—	—	—	270	—	—	—	—	—
Cocoa	288180	423589	565915	616812	889104	2764	3620	3282	4587	4749	0.959	0.855	0.580	0.744	0.534
Tobacco	298800	609800	673056	1154200	1088800	1098	1099	1297	2062	2108	0.367	0.180	0.193	0.179	0.194
Grain legumes (beans)	504136	523935	913097	1667100	1388570	2975	3726	4287	4662	4908	0.590	0.711	0.470	0.280	0.353
Bananas	600000	1051180	1473360	1857670	2963010	952	1358	1388	1772	2082	0.038	0.041	0.030	0.025	0.023
Plantain	1918130	2251380	3178340	5101820	6082110										
Cattle	13205720	14543200	18329420	16773310	18165494[a]	8717	9740	9515	14048	15652	0.066	0.067	0.052	0.084	—
Pigs	2219000	3510900	3318400	5517400	7476077[a]	2580	3404	2679	4875	9261	0.116	0.097	0.081	0.088	—
Sheep	30799	40340	54300	84110	92310	1887	2132	1960	2577	2929	6.127	5.285	3.610	3.064	3.173
Poultry	3582720	5001300	6820710	8577250	11476360[a]	2726	3386	2439	5684	5103	0.076	0.068	0.036	0.066	—
Minor species	—	—	—	—	—	—	—	100	387	1332	—	—	—	—	—
Forestry	927000	1216000	1770000	1950000	2668000	200	200	267	3937	4222	0.022	0.016	0.015	0.202	0.158
Fishery	915000	1036000	1598000	1920000	2534000	1608	11819	12503	15197	15294	0.176	1.141	0.782	0.792	0.604
Basic research						24361	29211	31053	42868	48116					
Support research						10319	14699	15746	20963	28412					
Operation research centres						68314	80203	96476	108520	121313					
Total	45264089	61493953	74239688	91348839	122036030	155776	206997	228605	293580	332300	0.344	0.337	0.308	0.321	0.272

[a] Estimated.

Source: IDRC. 1980. Project ARIAL. Asignación de recursos para investigación en América Latina. Colombia: estudio de caso.

Table 3. Relationship between total agricultural research expenditures and gross domestic product (total GDP and agricultural GDP) (thousands of Colombian pesos).

Year	Total agricultural research expenditure (A)	Total GDP (B)	Agricultural GDP (C)	A/B (%)	A/C (%)
1972	155776	186092300	49465000	0.08	0.32
1973	206997	243235900	66746000	0.09	0.31
1974	228605	329155400	88477600	0.07	0.26
1975	293580	412828700	113484800	0.07	0.26
1976	332300	532960800	148956300	0.06	0.22

Source: IDRC. 1980. Project ARIAL. Asignación de recursos para investigación en América Latina. Colombia: estudio de caso.

Table 4. Distribution of the ICA budget in different activities (millions of Colombian pesos).

Activity	1970	1971	1972	1973	1974	1975	1976	1977	1978
Administration	43.6	43.1	47.3	49.9	51.9	88.8	92.3	106.7	137.8
Debt service	—	0.1	1.9	6.8	12.4	28.6	60.6	63.9	73.8
Rural development	51.0	57.9	69.3	89.4	103.0	117.2	149.1	199.3	301.8
Research	121.3	143.6	151.2	175.5	188.1	236.7	266.7	307.8	420.4
	(121.3)[a]	(130.0)	(120.8)	(114.9)	(96.5)	(100.7)	(91.6)	(88.4)	(100.9)
Agricultural production	16.0	21.5	30.8	36.3	43.3	52.9	62.5	78.4	88.9
Livestock production	26.1	44.0	55.4	73.6	89.2	151.8	171.7	162.7	230.5
Physical investments and others	37.1	54.6	56.1	13.7	18.7	32.2	40.6	99.2	262.1
Total	295.1	364.8	412.0	445.2	506.6	708.2	843.5	1018.0	1515.3
	(295.1)	(332.2)	(329.1)	(291.6)	(259.8)	(301.4)	(298.7)	(292.2)	(363.7)

[a] Figures in parentheses are expressed in constant 1970 values.
Source: IDRC. 1980. Project ARIAL. Asignación de recursos para investigación en América Latina. Colombia: estudio de caso.

annual oleaginous products, sugarcane for sugar loaf (panela), cocoa, cotton, wheat, and tobacco; (3) low priority: plantains and bananas, sugarcane, barley, and oats.

Human Resources in Agricultural Research

General Trends in the Development of Human Resources

An ICA study[8] showed that the evolution of this institution's human resources has two main characteristics:

(1) In 1974, the research department of ICA had the highest concentration of university-trained professionals in the institution, either at the bachelor, M.S., or Ph.D. levels. By 1979, the relative importance of this department in terms of the number of professionals working in it had diminished (Table 10).

(2) Although most of the M.S. and Ph.D. holders working in the institute work in research, the percentage of them working in this area has been on the wane.

Brain Drain: Migration of Researchers[9]

Between 1960 and 1978, 652 persons were trained at the M.S. and Ph.D. levels. Of this group, 396 professionals were still working in ICA in 1978 and 256 had left. More importantly, the number of graduate level professionals who have left ICA has increased more rapidly than the number who have been hired.

---

[8] ICA. 1979. Diagnóstico de la investigación agropecuaria. Three volumes. (Unpublished).

[9] Based on the document: IICA. 1979. Sistemas nacionales de investigación agropecuaria en América Latina: análisis comparativo de los recursos humanos en países seleccionados. El caso del Instituto Colombiano Agropecuario (ICA). Volume III.2.

Table 5. Percentage participation of agricultural and livestock research in total research expenditures of ICA.[a]

| Year | Agricultural research | | | Livestock research | | | | Support research[b] |
	Research programs on crops	Basic research	Total	Program product	Basic research	Other	Total	
1972	41.3	17.6	58.9	19.2	9.5	4.2	32.9	8.3
1973	39.4	16.2	55.6	18.8	11.2	2.5	32.5	11.8
1974	40.0	18.1	58.1	16.3	10.9	3.8	31.0	11.0
1975	38.2	17.3	55.5	19.1	11.2	4.3	34.6	9.9
1976	36.3	16.2	52.5	19.8	11.3	5.0	36.1	11.4

[a] Does not include the operational costs of agricultural research stations.
[b] Includes biometry, agricultural resources, agricultural machinery, regional agricultural economy, etc.
Source: IDRC. 1980. Project ARIAL. Asignación de recursos para investigación en América Latina. Colombia: estudio de caso.

A recent study on the evolution of the human resources in ICA shows the following trend:[10]

	Researchers at graduate level (A)	Researchers leaving ICA (B)	B/A (%)
1960–67	63	2	3.2
1968–74	186	50	26.9
1975–78	104	55	52.9

Thus, there is a definite trend toward higher migration of researchers, coupled with less hiring of research staff. If this trend continues, the number of skilled researchers leaving the institute will outnumber those entering and ICA will suffer a net loss of highly trained graduate level staff.

Conclusions

This brief analysis of the situation of agricultural research in Colombia clearly points out three important trends that are having a negative impact on the sector:

(1) Funds allocated for agricultural research (both at the national level and in ICA) have been decreasing in real terms (in constant 1970 values) over the last decade (Tables 1, 4). This trend is also evident in the deterioration of the proportion of agricultural GDP that is allocated to agricultural research (Table 3).

(2) During the period under analysis, ICA has been increasingly assigned more duties but has not received a proportionate increase in budget funds. Consequently, the institute's departments compete for available resources. Research, formerly the most important ICA activity, has been negatively affected by this situation, both in terms of funds allocated to it within the ICA budget (Table 4) and in terms of

Table 6. Percentage participation of each crop in total research expenditure of ICA.

Crop	1972	1973	1974	1975	1976
Cereals	*13.0*	*13.2*	*14.5*	*11.5*	*11.1*
Rice	3.4	3.0	2.9	3.1	3.3
Oats	0.9	1.7	1.1	0.4	0.4
Barley	0.9	1.0	0.9	0.6	0.4
Maize and sorghum	6.0	5.6	6.8	4.8	4.6
Wheat	1.8	1.9	2.8	2.6	2.4
Starchy Crops	*6.1*	*5.6*	*5.6*	*5.5*	*5.4*
Potatoes and cassava	4.9	4.2	4.2	4.2	4.1
Plantain and bananas	1.2	1.4	1.4	1.3	1.3
Sugars	*1.3*	*1.3*	*1.1*	*2.5*	*2.9*
"Panela" (sugar loaf)	—	—	—	1.3	2.9
Sugarcane	1.3	1.3	1.1	1.2	—
Oil Seeds	*7.8*	*7.4*	*7.1*	*7.2*	*6.5*
Perennial	4.9	5.1	5.1	5.3	4.7
Cotton	2.9	2.3	2.0	1.9	1.8
Other Crops	*13.0*	*11.9*	*11.7*	*11.7*	*10.5*
Cocoa	3.3	3.6	3.2	3.3	2.9
Vegetables and fruits	5.0	4.0	3.5	4.0	3.7
Grain legumes and annual oil seeds	3.4	3.2	3.7	2.9	2.6
Tobacco	1.3	1.1	1.3	1.5	1.3
Total[a]	41.3	39.4	40.0	38.2	36.3

[a] Refers to the total percentage allocation to research programs on crops (Table 5).
Source: IDRC. 1980. Project ARIAL. Asignación de recursos para investigación en América Latina. Colombia: estudio de caso.

---

[10] IICA. 1979. Sistemas nacionales de investigación agropecuaria en América Latina: análisis comparativo de los recursos humanos en paises seleccionados. El caso del Instituto Colombiano Agropecuario (ICA). Volume III. 2, 36–38.

Table 7. Percentage participation of animal products in total research expenditures of ICA.

Animal program product	1972	1973	1974	1975	1976
Beef cattle	4.3	3.8	4.0	4.9	4.5
Dairy cattle	6.3	6.0	5.4	5.0	5.0
Pigs	3.1	3.4	2.5	3.2	5.4
Sheep	2.3	2.1	1.9	1.8	1.8
Poultry	3.3	3.4	2.4	4.0	2.9
Minor species	—	—	—	0.2	0.3
Total[a]	19.2	18.8	16.3	19.1	19.8

[a] Refers to the total percentage allocation to program-product livestock research (Table 5).
Source: IDRC. 1980. Project ARIAL. Asignación de recursos para investigación en América Latina. Colombia: estudio de caso.

Table 8. Percentage participation of main disciplines related to basic agricultural research in total research expenditure of ICA.

	1972	1973	1974	1975	1976
Entomology	3.4	2.1	3.2	2.6	2.9
Plant physiology	3.1	3.0	3.0	2.6	2.6
Plant pathology	4.4	4.5	4.9	4.5	4.2
Soils	6.7	6.6	7.1	7.6	6.4
Total[a]	17.6	16.2	18.1	17.3	16.2

[a] Refers to the total percentage allocation to basic agricultural research (Table 5).
Source: IDRC. 1980. Project ARIAL. Asignación de recursos para investigación en América Latina. Colombia: estudio de caso.

Table 9. Percentage participation of main disciplines related to basic livestock research in total research expenditure of ICA.

	1972	1973	1974	1975	1976
Animal physiology	0.8	0.9	1.2	1.0	1.0
Microbiology	3.6	4.5	4.5	3.9	4.1
Nutrition	0.8	0.9	1.0	0.8	0.7
Parasitology	1.5	2.0	1.5	1.8	1.1
Pathology	2.3	2.5	2.1	2.1	2.2
Toxicology	0.4	0.5	0.7	0.5	0.5
Epidemiology	—	—	—	0.2	0.4
Vascular diseases	—	—	—	1.1	1.3
Total[a]	9.5	11.2	10.9	11.2	11.3

[a] Refers to the total percentage allocation to basic livestock research (Table 5).
Source: IDRC. 1980. Project ARIAL. Asignación de recursos para investigación en América Latina. Colombia: estudio de caso.

high-level manpower dedicated to research in the institution (Table 10).

(3) Despite the effort made to train high-level manpower for research (M.S. and Ph.D. levels) carried out in the sixties and early seventies, ICA is facing an increasing problem of migration of researchers, coupled with less hiring and training of research staff. If this trend continues, its capacity to conduct research will be seriously impaired in the very near future.

It is in response to this deteriorating situation that the National Agricultural Research Plan was formulated. The plan is part of a broader package of government action aimed at changing the situation and stopping the downward trends. Two other important measures that form part of this package are the creation of a Special Fund for Agricultural Research (different from, and additional to, the ICA budget) and the establishment of a National Council for Agricultural Research and Technology Diffusion. These two measures are presently being considered in the Ministry of Agriculture and in Congress.

It should also be pointed out that the design and establishment of a Special Fund for Agricultural Research raises the important issue of identifying alternative financial mechanisms or systems for funding agricultural research within the country. The national budget has been the traditional source of research funds for this sector, given the centralized institutional model that has operated mainly around one large public research organization. For the creation of the special fund, alternative mechanisms for the mobilization of financial resources are being considered. This also raises the issue of the participation of the private sector in agricultural research and of mixed or joint research mechanisms between the public and private sectors.

A General Approach to the Process of Identifying Research Priorities in the Agricultural Sector

The formulation of research policies, in any field, is a way of responding to a situation in which multiple possible research topics compete for the limited financial resources that are available for supporting such activity. Furthermore, they are also a means for relating the research effort in any given country to the needs and development problems that are of major importance in that society. Research policies are also a means of influencing the characteristics and orientation of technical change and technological development in the agricultural sector, trying to make it more compatible with the "type of development" (or development objectives)

Table 10. Professional personnel by level of education in ICA.

	Bachelor's degree			M.S.			Ph.D.			Total		
Department	1974	1976	1979	1974	1976	1979	1974	1976	1979	1974	1976	1979
Research	406	205	137	77	155	145	34	32	39	517	392	321
Rural development	256	190	149	22	76	88	1	7	6	279	273	243
Livestock production	279	220	120	14	23	26	2	4	2	295	247	148
Agricultural products	120	95	42	8	26	24	2	3	2	130	124	68
Transfer of technology	—	—	222	—	—	17	—	—	—	—	—	239
Administration and planning	106	64	59	17	23	20	2	5	2	125	92	81
Total	1167	774	729	138	303	320	41	51	51	1346	1128	1100

Source: ICA. 1979. Diagnóstico de la investigación agropecuaria. Three volumes. (Unpublished).

that are considered to be most appropriate for that society. This third aspect leads to the broader issue of a "technological development policy" for the agricultural sector, of which the research policy is only one of several components. The orientation of technical change and technological development in the agricultural sector will depend, to a large extent, on a broad range of decisions that are made either by governments or by the producers themselves (at the level of the production units), such as decisions relating to what products should be produced in the country and which ones should be imported, what technologies should be made available or should be used, and what production systems should be promoted (i.e., cropping systems, size and type of production units, etc.). It is through these and other decisions that the "technological profile" of the agricultural sector will be determined and the dynamics of technical change will gradually take form.

Although the supply of technical knowledge generated by research programs is one of the factors that may influence these decisions (i.e., by making some alternatives possible or feasible), most often they are influenced by economic policies or market situations (both the national and international market) that confront the producer. Thus, many of the decisions are shaped by credit, commercialization, fiscal, monetary, and foreign exchange policies and foreign trade. These policies may also influence the relative importance that is given to national agricultural research efforts in any given period, and thus the financial resources that are allocated to agricultural activity. The role assigned to the agricultural sector in the development process by governmental policies (i.e., its relationship to industrialization and other developmental policies) also plays a major role. A preliminary analysis of the role played by some of these economic policies in Colombia is outlined later in this paper.

The previous considerations clearly point out that the agricultural research policy in any country is only one of the components of the technological development policy of that sector. This paper only addresses methodological issues related to the formulation of a research policy for the agricultural sector, with marginal references to the interphase of research policies with technological development considerations and economic policies that are of relevance to the sector.

At the most general level, research priorities can be derived from three major sources or considerations:

(1) Socioeconomic development policies and programs of a country, both at the global (i.e., general development programs, foreign trade policy) and sectoral levels (i.e., agricultural development policies, programs, and priorities). The objective is to link research efforts with the development objectives and priorities of a country.

(2) Specific needs or requirements that may be identified, both in terms of general needs of the country (i.e., the need to supply certain kinds of food for a specific sector of the population or the need to make better use of local food crops or natural resources) and specific requirements or problems related to agricultural production (i.e., the need to solve specific technological constraints that limit productivity in certain areas).

(3) Prospective considerations with respect to future agricultural needs, future expected situations of national and international agricultural markets, and the type of agricultural production system or food system one would like to develop in the future.

The importance of the first factor will depend on the existence of explicit and clearly defined agricultural development policies and programs in any given country. If these do not exist or if they are formulated only in vague and general terms (without specific priorities, development objectives, and production targets), as is quite often the case, this factor will play a smaller role in determining research priorities.

Nevertheless, even when explicit sectoral policies and development programs are clearly formulated, the criteria and guidelines derived from them should be complemented by the other two factors. The second factor may lead to the identification of requirements or production possibilities that are not adequately dealt with in the present sectoral development programs, such as the need to develop a "cropping systems" approach or the possibility of promoting greater use of traditional food crops existing in the country. If these requirements or possibilities are identified, they should be taken into consideration in order to correct possible gaps in the sectoral development plans.

Finally, both existing needs and sectoral development plans are normally conceived in terms of the present and very near future. Medium- and long-term perspectives are quite often absent from these considerations, or they play only a marginal role. The third factor is the most difficult to cope with, both in sectoral development planning efforts and in the identification and formulation of research priorities. The Colombian experience analyzed in this paper deals mainly with the first two factors. The prospective approach has not played a major role in this planning effort.

Methodological Framework for the Identification of Research Priorities

The formulation of a research policy for the agricultural sector involves three major levels of analysis:

(1) The identification of agricultural products or crops that have high socioeconomic importance or priority for the development of the country. The present or potential socioeconomic importance of certain crops is one of the criteria that may lead to the identification of research priorities but by itself does not define research priorities. Research areas are defined not only in terms of agricultural products or crops but also in terms of production problems or rural development issues, such as agricultural machinery and implements, irrigation technology and water supply, conservation and storage of crops, etc.

(2) Having identified agricultural products or crops that have high socioeconomic importance, the next step is to define which of them should receive major attention from the point of view of research. Given a situation of limited financial resources, not all products with present or potential socioeconomic importance can be covered by the research establishment of any country. This raises the following questions: Which products should be produced in the country and which should be imported? Which products face identifiable "technological constraints" that limit productivity and may lead to important research problems? Should the technology be generated internally (i.e., improving traditional or existing technologies) or should it simply be imported and adapted? Which products (research areas) should receive more support from government funds and which ones should be left to the initiative (and financial support) of the private sector? This last question is important in those countries where the private sector plays (or may play) a role in agricultural research. This second step narrows the range of products or production problems identified as important in the first step. Some of these questions imply political decisions (policy decisions).

(3) The third step consists of identifying or defining research topics or issues that are important for the solution of the technological constraints that limit production or productivity levels in the crops that have been selected. It is only in this third level of analysis that research priorities are actually formulated.

The preceding considerations define a general framework for the identification of research priorities and technological development objectives that is summarized in Fig. 1. The output of the socioeconomic considerations is the identification of (adjusted) socioeconomic product or problem priorities for research purposes.[11] The technological considerations of the process consist of the identification of technological requirements or problems within the selected products or problem areas that may lead to the identification of specific research needs (and, therefore, research priorities). The starting point for this analysis is the identification of the principal technological constraints that limit production or productivity levels of specific crops under identifiable circumstances. Technological constraints refer to physiological, environmental, or pathological factors, as well as management systems and farming practices, that are presently an obstacle to increasing production levels or improving the efficiency of resource utilization in specific crops or products (or even having a negative effect on these aspects).

The research effort that will have to be carried out in order to solve the technological constraints identified will depend not only on the socioeconomic

---

[11] It should again be emphasized that these may be slightly different from the priorities that may emerge from using only economic indicators.

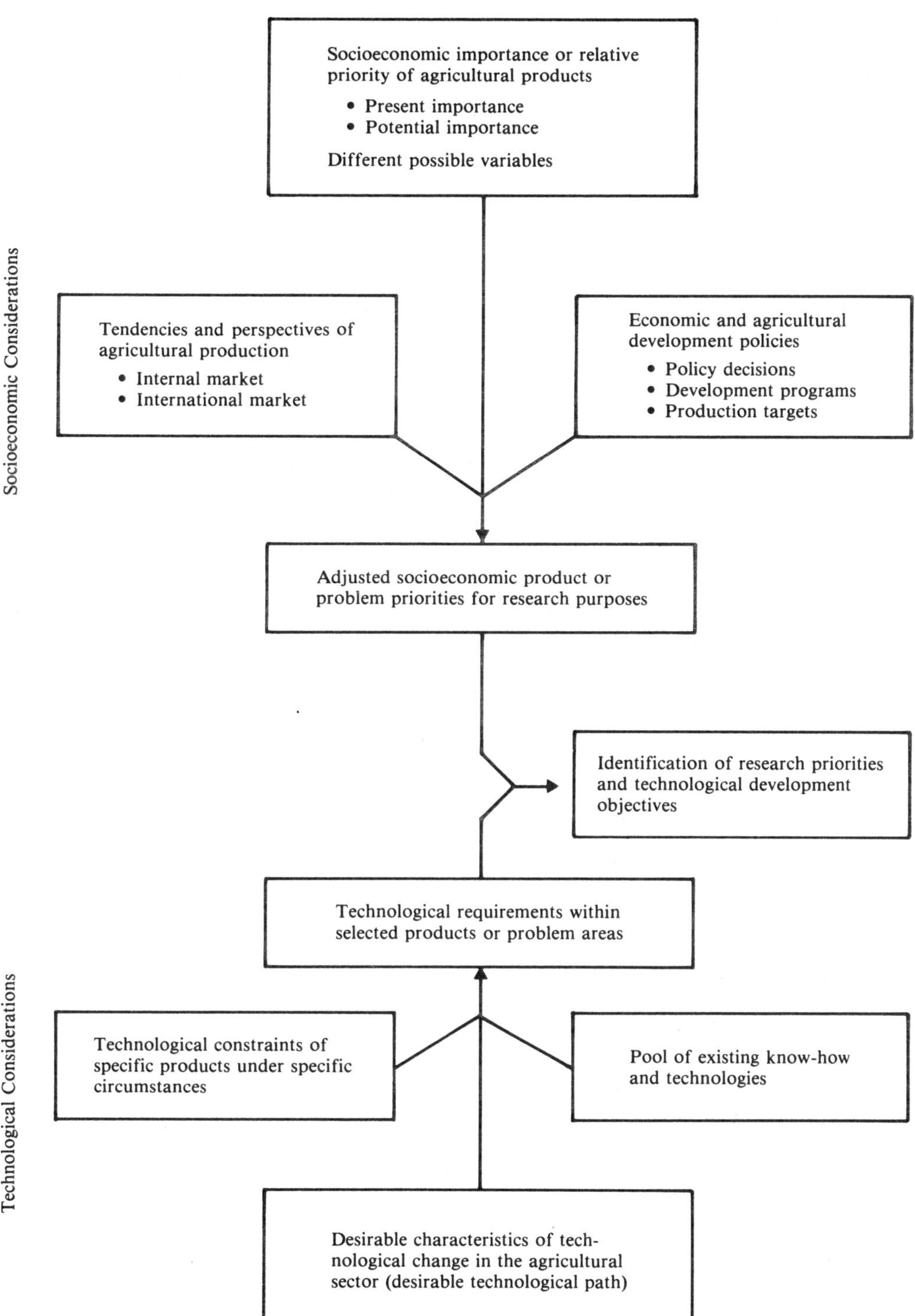

Fig. 1. Methodological framework for the identification of research priorities.

importance of the product but also on the difficulty or magnitude of the technological problem that is confronted. For example, in cases where the level of technological development (technological conditions of production) is considered to be acceptable for a specific crop in a given country, only research at a level necessary for maintaining existing high yields or disease-resisting varieties will be necessary (even for a high-priority crop). The research effort required (and the research priority) would be much higher, on the other hand, if important technological constraints are identified in a high-priority product. Thus, the order of product priority assigned on the basis of socioeconomic considerations can be altered or modified in view of technological considerations. It is for this reason that in Fig. 1, the final research priorities and the technological development objectives are derived from both types of considerations.

In Colombia, two analytical models are being simultaneously considered (and experimentally applied) in the process of defining socioeconomic product priorities for research purposes. These two models, although they can be used in a complementary manner, are based on a different set of variables or indicators for the identification of socioeconomic priorities.

The first model uses jointly, and tries to relate, two major criteria for priority identification: the comparative advantage a country has in producing a given crop and the participation of that crop in national food consumption or total family budget (argument of food security). Furthermore, this model uses the concept of price-demand elasticity to determine which products should receive higher priority in governmental support for research related to them and which product should be left to the initiative and funding of the private sector.

The second model uses as the main criteria the participation of each crop in the "total circulation of agricultural production" (this includes production for the internal market, exports, and imports of agricultural products). Besides these production variables, two additional indicators are taken into consideration to see if the model gains in analytical or discriminatory power (by substantially modifying the priorities initially identified). These are rural employment generated by each crop and the extension of land under a given crop's production.

The first model is more conceptually sophisticated and takes into consideration a broader range of factors, including major policy decisions that have to be made as part of the process of identifying priorities (i.e., export orientation versus food security and public versus private funding of agricultural research). On the other hand, however, it requires much more data, as well as the utilization of such concepts as "shadow prices" and the social costs of the use of domestic resources (land, capital, and labour).

The second model is much simpler and only requires data that is easy to use and readily available in any country. Its major assumption is that the participation of a crop in the total circulation of agricultural production has such close interrelationships with several other aspects or indicators of agricultural production (i.e., extension of land under that crop's production, total agricultural production, etc.) that it may be used as a significant approximation of socioeconomic importance or priority in terms of products. For example, the two additional variables to be discussed later do not add much to the priority ranking established by this basic criteria.

To identify technological requirements or problems within selected products or problem areas and derive research priorities from these requirements, ICA established a series of working groups covering the main crops that are produced in the country. The working methodology that was used has two main characteristics: (1) a matrix approach that tries to identify technological constraints on specific crops under certain environmental conditions that define ecologically homogeneous zones; in order to use this methodology the country was regionalized and divided into ecological zones; and (2) the use of the delphic technique, at the level of the different working groups, to identify and analyze the technological constraints and research needs that are faced by each crop.

The output of this process has been the formulation of "research programs" for the different crops or agricultural products under consideration. The set of research programs thus formulated, with a few other components related to general policy issues, constitute the "National Plan for Agricultural Research."

Some Observations with Respect to the Application of this Methodological Framework in Colombia

It should be noted that the two main phases of this planning process (i.e., the identification of socioeconomic product or problem priorities and the determination of technological requirements and research needs within selected products or problem areas) are supposed to be carried out in chronological order, i.e., the determination of technological requirements and research needs within products or problem areas should be carried out only for those products and problems identified as having high (or significant) socioeconomic importance for the country. This, of course, implies that the policy decisions that are raised by the two models have been coped with and answered.

Nevertheless, the sequence of events in real life situations does not always follow the logical ordering of methodological steps. In fact, the two phases of this planning process may overlap and be carried out simultaneously or in parallel fashion, as in Colombia. In this case, ICA decided to go into the identification of technological requirements and research needs at the product level (second phase), although there was still much ongoing discussion as to which were the agricultural product and problem areas that could be considered to have high socioeconomic importance. The two models were developed in response to this issue, but even though the first phase is still an ongoing process in Colombia (the two models are being experimentally applied), ICA has already finished formulating a first version of the research programs that should be carried out at the level of each product. Thus, the methodology of the second phase has already been tried out and empirically tested, having reached the stage of producing a first version of possible research programs at the product level.[12]

The analysis of the reasons for the discrepancy between the methodological framework or approach that has been presented and its actual implementation in Colombia gives an interesting insight into the dynamics of the planning process and into some of the practical problems that it faces.

When this planning process started, it soon became clear that although the determination of technological requirements and research needs within products (second phase) was, basically, a technical endeavour, which could be easily implemented if the necessary information was available, the identification of socioeconomic priorities (first phase) involved policy decisions with respect to the criteria (model) to be used and with respect to substantive economic policy issues. This being the case, the decision-making process with respect to the latter component proved to be much slower and more difficult than had been expected. Consequently, it took some time to develop and discuss the two models that are presented.[13]

In order not to stop the process of identifying research priorities and formulating research programs at the product level, until the basic issue of defining socioeconomic priorities was settled (which could become a vicious circle), an alternative route was taken. It was decided to use a list of 28 products that the Ministry of Agriculture (OPSA) had drawn up, which represents almost the total agricultural production within the country and for which there is information on production and commercialization. In fact, the 28 products represent 97% of total agricultural production. The process of identifying technological requirements and research needs within specific products (second phase) was carried out for all 28 products.

The implications of this operational decision are obvious. Because the 28 products do not reflect any evaluation of socioeconomic priority (it is merely a list of the products that are being produced in the country), the proposed research programs cover almost the total range of agricultural production and, therefore, the total range of possible research topics in terms of products.[14]

Despite this limitation, the alternative was adopted for the following reasons:

(1) The procedure does not invalidate the effort of identifying technological constraints and research needs within products (second phase). It merely made it a more manpower-intensive and costly process because the exercise was carried out not only for high-priority products but for almost all products. On the other hand, however, it was considered that this planning exercise would produce valuable information on technological constraints and problems that are faced by agricultural production within the country (even for low-priority products).

(2) Because socioeconomic product priorities have not yet been established due to the difficulties encountered in the first phase), the first version of the National Plan for Agricultural Research suggests a resource-allocation procedure (and thus implicit priorities) in terms of the relative importance of each product from the point of view of its participation in the total agricultural production at the present time, and in terms of the need to create a basic research infrastructure in some research areas (requiring higher investment levels).

(3) It was considered that the result of establishing explicit socioeconomic product priorities (once the first phase of this methodological process is completed) could be incorporated a posteriori into the final version of the National Plan for Agricultural Research by modifying, accordingly, the respective importance given to the different research programs for resource-allocation purposes and, if necessary,

---

[12] See Plan nacional de investigación agropecuaria del ICA. 1981. Five volumes.

[13] Because the two models were only recently developed, final policy decisions with respect to the priorities that emerge from them in the Colombian case are still pending.

[14] With the exception of coffee and sugarcane, which in the case of Colombia are research areas that are in the hands of the private sector.

by eliminating those programs for low-priority products.

Thus, the Colombian experience shows complex interaction between the two major phases of the methodological framework presented, given the need to adapt formal procedures and methodological steps to the realities and conditions of the planning process within each country.

Identification of Socioeconomic Priorities in Terms of Products

Identification of Socioeconomic Priorities in Terms of Comparative Advantages and Food Security

The theory of induced technological change, endogenous to the economic system, suggests that the relative price of factors affects both the choice of existing technology as well as the biases in the use of factors in the new production functions. It has been shown empirically that the different paths of technological development taken by the United States and Japan have been determined by the relative price of factors that reflect the different endowments these countries have in terms of land and labour.

In underdeveloped countries, it has been found that when governments establish the price of goods and factors without taking into account a country's endowment of factors, patterns of technological change are not compatible with a country's comparative advantages. In many developing countries, government policies undervalue certain kinds of products and overvalue others; the result is that errors are made in allocating resources for production.

Current economic theory has yet to explain why government makes this type of error in decision-making. Of course, government leaders have political commitments and the measures they take are politically motivated. The advocation of specific types of policy fundamentally depends upon the advantage political groups hope to gain from them. Thus, a ruling political group can impose its point of view and implement price policies and technological strategies that are incongruent with a country's particular endowment of factors.

Thus, political considerations filter down to decision-making levels where resources for research are allocated; these influences can significantly distort the process. Therefore, the evolution of overall development policies, especially those policies related to agriculture, must be considered when trying to find an explanation of how funds for agricultural research are allocated.

Economics has assigned agriculture certain functions in the economic development process. They include: (1) increasing the available food supply and freeing the labour force to work in nonagricultural sectors; (2) expanding the available market for industrial products; (3) increasing domestic savings; and (4) providing foreign exchange through agricultural exports.

The analysis of closed economies generally contains the first three points. However, when dealing with an open economy and when confronted with the fourth point, the other points no longer relate to domestic agriculture alone and could even become incompatible. The concept of comparative advantage is the relevant concept, in open economies, to evaluate efficiency or inefficiency in the allocation of resources. For example, in an open economy, it is not always desirable for a country to produce its own foodstuffs if the food could be acquired more cheaply in international markets. Therefore, the nutritional importance of a product, or other similar yardsticks, does not provide a basis for assessing the efficiency with which resources are allocated for research unless other criteria are considered such as international prices and the cost of the domestic resources needed to produce the same product; this includes knowledge of the opportunity costs of capital, labour, land, and foreign exchange. The concept of social costs of production and factors becomes important when considering economies that are riddled with distortions. For example, the market price of a product often does not represent its true social value; therefore, a person allocating resources on the basis of the production value alone can over- or underallocate resources; this will depend upon a country's current pricing policy, i.e., whether a specific product is under- or overvalued. This, in turn, depends on the priorities of the party in power.

A country may decide to ignore these considerations for political reasons or because it does not want to take risks and decides to guarantee the availability of food. Consequently, the country might allocate large quantities of resources for products that are important for the nutrition of its inhabitants. This means that at a given point in time the country in question does not have enough confidence in its ability to purchase the amount of foodstuffs it requires on the international market in order to avoid sharp fluctuations in domestic supply, or that even though a country has sufficient foreign exchange, it views food availability as essential to defending itself from outside political pressures.

The approach proposed here is one of an open economy in which the allocation of resources for research is based on comparative advantages and guaranteed availability of food or self-sufficiency in terms of the world market. This approach also allows for the distortions within an economy (subsidized credit, minimum wage, tariffs, subsidies, etc.) that fundamentally influence the way resources are spent. Special emphasis is placed on the repercussions of the macroeconomic policies and development model a government adopts on agriculture in general, and on the process of generating and adopting technological change in particular.

The Influence of Economic Policies on Agricultural Research Trends

The Colombian experience shows that agricultural policy, and technological policy as a subdivision of this policy, are determined in the long term by the development policies and models adopted by the government and are defined in the short and medium terms by the evolution of certain important macroeconomic aggregates.

During the period of rapid industrialization between 1950 and 1967, Colombia followed the import-substitution model, which tried to protect domestic production by establishing high tariffs and import quotas on consumer goods. Overvaluing the peso was another key tool in this policy and constituted, in effect, a tax on exports (primarily agricultural exports). During the 1960s, when the bias toward substituting imports grew stronger, taxes on agricultural exports ran from 17–47%. Another means of subsidizing industrialization was to force farmers to sell raw materials such as cotton to domestic producers at lower than international prices. In the short term, such measures acted to discourage the production of these goods, and over the long term, they inhibited the generation and adoption of technology. Only those products for which the country had a true comparative advantage, such as coffee, sugarcane, tobacco, and cotton, could withstand the pressures of this model.

At the same time, this model of rapid industrialization created the need for a large work force, which received stable or declining real wages. A major part of this salary is spent on food, so the model requires that there be an abundant supply of fundamental foodstuffs. The limited foreign exchange generated by the economy must be spent on importing the intermediate and capital goods necessary to boost the industrial process. Foreign exchange cannot be spent on importing food and agricultural raw materials. Therefore, credit, prices, and research policies for this period stressed the production of certain foodstuffs and the import substitution of certain raw materials.

In 1967, the import-substitution model gave way to the promotion of exports; trade policy and the exchange rate immediately reflected this situation. From 1970 onward, exports increased considerably, and higher international prices for coffee produced more foreign exchange and a relatively large surplus in the balance of payments. This situation brought about a change of priorities in the allocation of resources. First, the importance of products that had substituted for imports declined; more wheat, corn, sorghum, oil, and milk were purchased abroad. However, the excess amount of foreign exchange and its resulting monetization quickened the pace of inflation in Colombia and favoured stabilization policies in the short term, so that food imports became increasingly necessary. The energy crisis occurred during this same period and more money was spent to explore for new sources of oil and to develop alternate sources of energy (hydroelectric, nuclear, etc.). All of these activities demanded large amounts of resources from the national budget.

As a consequence of this, in the mid-1970s, the government was not only forced to curtail public spending to stabilize the budget, but most available resources went toward solving the energy crisis. In addition, with a surplus in the balance of payments, the government did not seek out foreign credit to finance research efforts. This brief description of the Colombian situation and its trade, fiscal, monetary, and exchange policies helps explain why, during certain periods, the level of resources earmarked for agricultural research decreased. It also helps explain why, at a given time, large quantities of resources flow toward certain types of products.

A Model for Identifying Product Priorities: Comparative Advantages and Food Security

When setting priorities among products for the allocation of research resources, several fundamental points must be considered: characteristics of the country's production system — relative availability of land, labour, capital, foreign exchange, and the social costs of each of these factors; availability of food and raw materials to meet nutritional needs and the country's industrial production needs; overall development models and policies; and financial resources available for agricultural research.

Because an open economy framework is being used and because one of the priorities of the Colombian Development Plan is to generate a stable flow of foreign exchange (anticipating later balance of payment problems), a basic criterion that must be used when allocating resources for research is the concept of comparative advantage. When a country has a comparative advantage in the production of a commodity, the net social return on producing an additional unit of the product is positive. In other words, the value of the product in terms of its shadow price (for marketable products, the border price, CIF, or FOB) should be higher than the social cost of the resource earmarked for its production.[15]

The comparative advantage can be calculated by using a parameter known as the domestic resources cost (DRC). It measures the social cost, in terms of domestic resources (land, labour, capital), of generating one additional unit of foreign exchange either by exporting or by substituting imports. This cost is then compared with the average cost in the economy of generating the same unit of foreign exchange (shadow exchange rate); if the quotient is less than 1, the country has a comparative advantage in this area.[16] For example, in 1978, it was estimated that the shadow exchange rate for Colombia was 36 pesos to the U.S. dollar. However, the domestic resources cost to substitute one dollar in maize imports was 45 pesos. In this case, Colombia did not have a comparative advantage in maize production.[17]

Using the cost structure of the different products and the percentage of imported inputs for these products, it is simple to calculate the DRC and the comparative advantage; this makes it possible to work out a scale that orders products according to their comparative advantage, using 1 as the dividing point.

Nevertheless, considerations of comparative advantages cannot be used as the only criterion for resource allocation. It is necessary to combine this criterion with food self-sufficiency or guaranteed food supply. This is especially important because the National Development Plan in Colombia places great emphasis on generating a sufficient supply of food for the people, as well as providing sufficient raw materials for agroindustry.

In order to be able to use the argument of food self-sufficiency as a criterion for setting product priorities, it is necessary to establish the weight (participation rate) that each product has in the total family budget. This is an indicator of their importance in terms of the food supply that has to be guaranteed in the country. For agricultural products used as raw materials in industrial processes (i.e., soybeans for oil), this information can be estimated by establishing the agricultural product's share in the cost structure of the industrial product, and multiplying this percentage by the industrial product's share in the total family budget.[18]

On the basis of these two criteria, it is possible to set up a graph of priorities. Comparative advantage will run along the horizontal axis and the importance of a product in family spending along the vertical axis (Fig. 2). The products in quadrants I and IV are those in which the country has a comparative advantage, and can export or substitute for imports efficiently. The products in quadrant IV, due to their low position in family spending, are the easiest to export. The products in quadrant I make up a significant part of the consumer's shopping basket, in addition to the comparative advantage the country has in their production. Therefore, quadrant I contains products that could efficiently substitute for imports or could be exported. The products in quadrant II have no comparative advantage but make up a significant part of the consumer shopping basket. The social return on the resources invested in promoting their production is low; this also holds true for the products in quadrant III, whose share in family spending is low. The products in quadrant II are importable or potentially importable. Quadrant III shows importable and domestic products whose share in family spending is not high.

The highest research priority should be given to the products in quadrant I because they have a comparative advantage (RSN > 0); they are also key items in the consumer's shopping basket. The products in quadrant III have the lowest priority. Government policy definition would provide the information necessary to establish the difference between quadrants II and IV. If the government de-

---

[15] The social return on a specific activity can be measured using the following formula:

$$\text{RSN}_j = \sum_{i=1}^{n} a_{ij}\, P_i - \sum_{s=1}^{m} F_{sj}\, V_s + E_j$$

where: a_{ij} = amount of the ith product produced by activity j; P_i = shadow price of this product; F_{sj} = amount of sth production factor used by j; V_s = social cost of the sth factor; and E_j = external effect produced by activity j.

[16] The domestic resources cost can be calculated using the following formula:

$$\text{DRC}_j = \left(\sum_{s=2}^{m} F_{sj}\, V_s - E_j \right) \Big/ \text{VAN}_j = \text{CD}_j \Big/ \text{VAN}_j$$

where: CD_j = domestic opportunity cost of the resources used in j; and VAN = net foreign exchange earned or value added to international prices.

[17] The shadow exchange rate represents the average cost to the economy to produce one additional unit of foreign exchange.

[18] The products that are most difficult to classify are those used as raw materials in different industrial processes. Some products, such as cotton, are especially difficult because they are used in several processes (cotton is used in textiles and cottonseed cake); in such cases, one would have to choose the processes that occupy the most important place in family spending and on the basis of this percentage, estimate cotton's share in this spending.

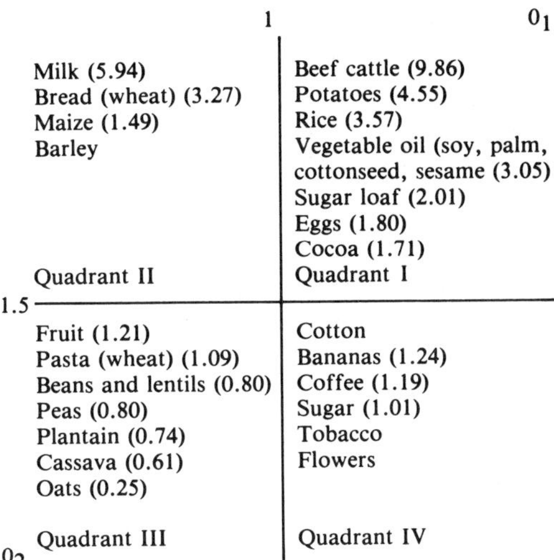

Fig. 2. Priorities of products using socioeconomic criteria. Point 0_1 represents the origin for comparative advantage or the quotient between the domestic cost of resources and the shadow exchange rate; point 0_2 represents the origin for the product's participation in family spending and is measured vertically. This participation or share of spending is shown in parentheses and represents the structure of spending for blue-collar workers in the city of Bogota. Comparative advantages are positioned subjectively and will remain so until the corresponding calculations have been made. Another way of situating along the vertical axis would use the quotient domestic production and consumption with a dividing line at point 1. This line would be the "line of self-sufficiency." In this case, point 0_2 will be at the top corner of the matrix.

cides to adopt a policy of promoting exports and obtaining foreign exchange to provide guaranteed supplies of food, quadrant IV would be favoured. However, if the government adopts a food self-sufficiency policy, quadrant II is favoured. Exporting countries adopting the first type of policy would prefer quadrants I and IV, whereas self-sufficient countries would choose quadrants I and II.

Furthermore, the products that should receive priority government financing and those that should be left to the initiative of the private sector must also be determined. This is done by examining the price elasticity of demand. When the demand for a product is inelastic, consumers reap the benefits of research; when demand is elastic, producers benefit from research. Therefore, the government should finance research on priority products having the least price elasticity of demand and continue up the scale until available resources are exhausted. Research on other products should be financed by the private

sector. Because exportable products usually have high price elasticity of demand, the products in quadrant IV would be financed by the private sector (coffee, sugarcane, cotton, etc.), whereas the government should handle the products in quadrants I and II. In Colombia, the choice of the products in quadrants I and IV would give products from the tropical zone a clear advantage over those from the Andean zone (except for coffee). Having set product priorities at the economic level, technological and research priorities must now be established.

Identification of Socioeconomic Priorities in Terms of the Internal and External Market for Agricultural Production

The Concept of Total Value of Agricultural Circulation

Among the main functions assigned to the agricultural sector in the economic development process, two aspects are of particular importance: the satisfaction of the internal demand for food and raw materials needed in the industrial sector (production for the internal market); and the generation of foreign exchange needed to sustain the development of the national production system, both through agricultural exports and through the substitution of agricultural imports (exports and imports). These two aspects are of central importance to some of the other functions assigned to this sector, such as the broadening of the domestic market for goods and services produced in other sectors of the economy, and the liberation of part of the labour force to work in nonagricultural activities.

The capacity of the agricultural sector to carry out these functions depends, to a large extent, on the magnitude of the gross agricultural product generated by the sector. It is for this reason that one of the most common indicators to measure the relative importance of each agricultural product, in terms of the function it performs within the whole economy, has been the participation of that product in the total value of agricultural production.

In order to take into consideration the different functions that have been assigned to the agricultural sector, a more appropriate indicator appears to be the total value generated by the circulation of agricultural products in a given economy, which will be referred to as the total value of agricultural circulation.

The value generated by the circulation of agricultural products has three major components or sources: agricultural production for the internal market (APIM); agricultural exports (X); and agricultural imports (M). The total value of agricultural circulation (AC) is defined as the sum of the value generated by these three components, i.e., AC = APIM + X + M. This indicator, which is somewhat different from that of the total value of agricultural production, takes into consideration the three dimensions that were identified with respect to the main functions assigned to the agricultural sector in the process of economic development, i.e., production for the internal market (satisfaction of the demand for food and raw materials), agricultural exports, and agricultural imports.[19] Table 11 shows the total value of agricultural circulation in Colombia from 1972–1976, as well as the annual value of its three components (in constant values of 1970).

A Model for the Identification of Product Priorities: Participation in the Total Value of Agricultural Circulation

The basic premise of this model is that the relative importance of every agricultural product, in terms of the function it performs within the whole economy, can be established on the basis of the participation of that product in the total value of agricultural circulation. A "general priority index" for each crop or agricultural product can be computed through the following procedure:

(1) The first step is to determine the total value of agricultural circulation in the country during a given time period. This entails: (a) Disaggregation of the total value of agricultural production into its two major components, production for the internal market and agricultural exports. The value of production for the internal market is estimated on the basis of producer's prices; production for agricultural exports is established by converting the FOB value of exports into local currency. (b) The value of agricultural imports (at CIF prices) is converted into local currency.

(2) The relative importance of these three components is established in terms of their percentage participation in the total value of agricultural circulation. This is done not only on the basis of a single year, but on the basis of the average annual value over a number of years, in order to avoid distortions of exceptional exports or imports in any given year. Thus, Table 11 indicates the annual values of these three components in Colombia from 1972–1976, as well as the average annual value of them for this time period. This last information indicates that in Colombia, production for the internal market represents 71.6% of the total value of agricultural circulation, whereas exports represent 25.3% and imports constitute only 3.1% of the total value. These three percentages are used as "weighting coefficients or parameters" in a subsequent step of this method.

(3) The percentage participation of each crop or agricultural product in the three components under analysis is determined. This provides information with respect to the relative importance of each product in agricultural production for the internal market, agricultural exports, and agricultural imports. Information related to Colombia is given in Tables 12, 13 and 14 for the 1972–1976 period.

(4) The "general priority index" for each crop or agricultural product can be computed as follows: (a) The percentage participation of each crop making up the three components of agricultural circulation is multiplied by the relative importance or weight of the respective component in the total value of agricultural circulation. This weighting procedure, which uses the coefficients determined in the second step, provides the "weighted participation" of the different crops in the three components of agricultural circulation. (b) The "general priority index" for each crop is computed by adding the "weighted coefficients of participation" of that crop in the three components under analysis. It should be noted that normally any given crop appears in two of the three components, because it is only under very special circumstances that the same crop is both exported and imported in a specific country.

The procedure can be better understood through an example. As seen in Table 15, the percentage participation of coffee in the components of agricultural circulation in Colombia is: in production for the domestic market, 5%; in exports, 73.3%; in imports, 0%. Because the relative weight of each of these components in the Colombian case is 71.6%, 25.3%, and 3.1%, respectively, the weighting procedure described above and the general priority index of coffee in this country is:

	Partici-pation (%)	Weighting coefficient	Weighted partici-pation
Internal market	5.0	71.6	3.58
Exports	73.3	25.3	18.54
Imports	0.0	3.1	0.00
General priority index			22.12

Quantitative indicators of relative priorities can be effectively used as one of the main criteria in the final decision-making process for resource allocation but they should not be considered as the only criteria. At least two other aspects should be taken into consideration. In the first place, as a result of a

---

political decision, and aside from any considerations on social returns, it could be decided to stimulate certain products as part of a national policy of guaranteeing the internal supply of those food crops or raw materials. Secondly, an analysis of past production trends and the future outlook for certain crops may identify agricultural products of potential importance to the country, although specific crops may not be of major importance in terms of present levels of production. This may be the case for some minor or nontraditional crops in any given country. Thus, the priorities established should be partially modified or adjusted in the final decision-making process. Nevertheless, this does not invalidate the indicators and procedure that have been presented because they provide a clear basis for decision-making in the process of resource allocation for agricultural research.

It should also be pointed out that in the application of this model in Colombia, two additional variables or indicators were taken into account to see if the model gained in analytical power by substantially modifying the priorities initially identified. These additional variables were rural employment generated by each crop and the extension of land (area) under that crop's production. No significant modification was introduced by these variables in the priority ranking established on the basis of the indicators that have been suggested.

A final methodological note is in order with respect to the choice of shadow prices versus market prices in analyzing the three components of the total value of agricultural circulation. In the application of this model in Colombia, market prices were chosen for two reasons. Firstly, the price of most agricultural products in Colombia is not substantially distorted by political and institutional actions; thus, the difference between market prices and shadow prices is not considered to be significant. If this were not the case, the use of shadow prices might be advisable. Secondly, due to the operational (data gathering) and conceptual difficulties related to the use of shadow prices, it was felt that the additional precision to be gained by their use (in terms of a different and better priority ranking) would be so marginal that it would not compensate for the additional effort required in data gathering and processing.

One of the greatest operational advantages of the model presented is that the data it requires are readily available in any country and the application of the data entails no great difficulty. The observations made earlier, however, regarding the need to adjust the priority ranking established by the indicators that have been suggested, on the basis of political considerations or trend analysis, should be kept in mind.

Application of the Model to Colombia

This method for identifying product priorities for research purposes was applied to the 28 agricultural products that constitute most of the agricultural production in Colombia. Between 1972 and 1976 the annual average of the total value of agricultural circulation generated by this sector in Colombia was $47 139.3 million Colombian pesos (expressed in constant 1970 pesos). Of this total, production for the internal market represents 71.6%; agricultural exports represent 25.3%; and agricultural imports constitute the other 3.1% (Table 11). The annual average values over a number of years were used to avoid distortions that could be introduced by exceptional agricultural exports or imports in any given year.

Following the methodology previously described, the percentage participation of each crop or agricultural product in the three components of the total value of agricultural circulation was determined. Table 12 shows the percentage participation of the main agricultural products of the country in agricultural production for the internal market (1972–1976); Tables 13 and 14 show the relevant participation coefficients of these same products with respect to the value of agricultural exports (1972–1978) and imports (1972–1977) respectively. As in the previous case, an average annual participation rate of the different products, during a given time period, was computed in order to avoid the distortions that could be introduced by exceptionally high or low crops of a specific product in any given year.

On the basis of the information provided in Tables 11–14, the weighted participation coefficient and general priority index of each product were computed. The weighted participation coefficients of the main agricultural products of Colombia are shown in Table 15, as well as the general priority index of each product. This index measures the relative importance (or participation) of each product in the total value of agricultural circulation in the country during the time period being analyzed (1972–1976). The initial participation rates appearing in Table 15 are really average annual participation rates for this period, derived from Tables 12, 13, and 14.

For comparative purposes, Table 15 also includes information regarding the participation rates of the different crops and products in the total value of agricultural production for this same period. By comparing this with the general priority index, one can compare the priority rankings that are established by using participation rates in the total value

Table 11. Total value and structure of agricultural circulation, 1972–1976 (millions of constant 1970 pesos).

	1972	1973	1974	1975	1976	Average value	Weighting coefficients
Value of production for the internal market	31186.1	32047.5	34908.0	36472.2	34276.5	33778.1	71.6
Value of exports	10293.7	11947.4	11404.8	13467.5	12489.4	11920.6	25.3
Value of imports	741.1	1332.9	1906.1	1183.0	2040.6	1440.7	3.1
Total value of agricultural circulation	42220.9	45327.8	48218.9	51122.7	48806.5	47139.4	100.0

Source: Balcazar, Alvaro and Torres, Ricardo. 1981. Selección de prioridades socio-económicas para la investigación agropecuaria. COLCIENCIAS, p. 79.

Table 12. Percentage participation of main products in agricultural production for the internal market, 1972–1976.

Product	1972	1973	1974	1975	1976	Average
Coffee	5.4	5.0	4.9	3.8	5.7	5.0
Rice	4.8	7.7	8.3	7.0	6.1	6.8
Barley	0.5	0.5	0.5	0.8	0.4	0.5
Maize	4.5	5.0	3.9	3.4	4.3	4.2
Sorghum	1.1	1.6	1.6	1.4	1.7	1.5
Wheat	0.4	0.4	0.4	0.3	0.3	0.4
Potato	3.0	3.3	3.3	6.1	4.4	4.0
Plantain	4.9	4.6	4.7	5.9	6.1	5.2
Cassava	7.5	5.4	6.7	7.6	6.1	6.7
Yam	—	0.4	0.3	0.3	0.3	0.3
Sugarcane	3.2	2.9	2.6	2.4	4.4	3.1
"Panela"[a]	5.1	5.7	3.7	3.1	7.6	5.0
Soybean	0.8	0.8	1.0	1.4	0.6	0.9
African Palm	—	0.8	1.1	0.7	0.7	0.8
Sesame	0.4	0.2	0.3	0.3	0.3	0.3
Cotton	4.7	5.3	5.4	4.0	6.2	5.1
Cocoa	0.7	0.8	0.8	0.7	0.9	0.8
Tobacco	0.5	0.7	0.5	1.1	0.5	0.7
Beans	1.1	0.9	1.1	1.7	1.1	1.2
Bananas	1.0	1.4	1.2	1.0	1.6	1.2
Livestock						
Dairy	15.4	9.7	9.5	7.4	—	10.5
Beef	—	19.0	16.8	11.8	14.3	15.5
Pigs	5.7	7.2	4.9	6.4	—	6.0
Sheep	0.1	0.1	0.1	0.1	0.1	0.1
Poultry						
Meat	4.4	4.9	5.2	5.7	—	5.0
Eggs	4.7	5.3	4.8	4.3	4.5	4.7
Others	20.1[b]	0.4	6.4	11.3	21.8[b]	12.0

[a] Sugar loaf.

[b] The high unexplained percentages in these 2 years are due to the lack of information, in those particular years, for one or two important products.

Source: Balcazar, Alvaro and Torres, Ricardo. 1981. Selección de prioridades socio-económicas para la investigación agropecuaria. COLCIENCIAS, p. 75.

of agricultural production and participation rates in the total value of agricultural circulation. The difference between these two priority rankings is greater in those countries or products where agricultural imports play a more important role.[20] Thus, the difference is greater in products such as wheat in the Colombian case, due to the significant import component for this crop.

[20] The overall importance of agricultural imports in Colombia is not very significant, representing only 3.1% of the total value of agricultural circulation.

Table 13. Percentage participation of main products in the value of agricultural exports in Colombia, 1972–1978.

Product	1972	1973	1974	1975	1976	1977	1978	Average
Coffee	72.9	77.4	73.1	65.7	77.5	81.8	83.0	73.3
Bananas	2.3	2.0	3.0	3.1	3.7	2.5	3.1	2.8
Sugar	4.9	3.9	8.5	9.3	2.2	0.1	0.9	5.8
Cotton	8.7	4.3	5.6	7.4	5.3	6.3	3.0	6.2
Tobacco	1.7	2.0	2.1	1.2	2.3	1.1	1.1	1.9
Rice	0.1	0.1	—	2.0	1.9	1.1	1.1	0.8
Potatoes	—	—	—	0.2	0.1	0.1	0.1	—
Cocoa	—	—	—	—	—	—	—	—
Maize	—	—	—	0.2	—	—	—	—
Beans	0.3	0.3	1.0	0.8	0.5	0.5	—	0.6
Vegetables and legumes	—	—	—	—	—	—	—	—
Tomatoes	—	—	—	—	—	—	—	—
Soybeans	0.1	0.1	—	—	—	—	—	—
Oats	—	—	—	—	—	—	—	—
Flowers	0.5	1.1	1.9	1.9	2.0	1.8	2.2	1.5
Bovine stock	2.3	0.3	0.3	2.6	1.3	0.6	0.6	1.3
Beef cattle	4.1	5.2	3.8	2.2	1.8	1.3	2.0	3.4
Others	2.1	3.3	0.7	3.4	1.4	2.8	2.9	2.4

Source: Balcazar, Alvaro and Torres, Ricardo. 1981. Selección de prioridades socio-económicas para la investigación agropecuaria. COLCIENCIAS, p. 68.

Table 14. Percentage participation of main products in the value of agricultural imports, 1972–1977.

Product	1972	1973	1974	1975	1976	1977	Average 1972–76
Wheat	67.1	39.4	55.5	60.0	38.0	15.6	52.0
Maize	0.2	11.6	4.3	—	1.3	7.8	3.5
Beans	0.5	0.2	0.2	0.3	—	1.0	0.2
Barley	—	7.5	5.2	2.7	5.7	8.7	4.2
Soybean	2.9	7.4	6.4	—	—	—	3.3
Soybean oil	0.2	1.0	3.4	2.4	8.4	13.4	3.1
Peas	0.1	1.0	1.2	3.0	1.2	3.2	1.3
Chick-pea	0.4	2.8	1.0	—	0.3	0.5	0.9
Lentils	1.8	3.8	3.9	3.4	3.7	3.6	3.3
Apples	5.5	1.8	2.2	3.4	2.8	2.1	3.1
Oats	3.5	2.5	2.0	2.6	1.6	1.6	2.4
Cocoa	12.0	8.8	7.1	6.5	0.3	—	6.9
Beef cattle	0.1	0.1	0.2	0.2	0.2	0.5	0.1
Dairy cattle	1.9	1.5	1.4	1.8	4.8	12.0	2.3
Poultry	1.2	0.8	0.5	0.8	0.6	0.6	0.8
Eggs	0.1	—	—	—	—	—	—
Others	2.5	9.8	5.5	12.9	31.1	29.4	12.4

Source: Balcazar, Alvaro and Torres, Ricardo. 1981. Selección de prioridades socio-económicas para la investigactión agropecuaria. COLCIENCIAS, p. 73.

In addition to establishing a rank order among the 28 agricultural products being considered, the priority index can be used to identify clusters or groups of products upon which it is possible to classify the different products in terms of general priority levels: high, medium, and low priority. An analysis of the index in Table 15 identifies four groups of products:[21] group 1 (index value >7): coffee, beef

[21] The index values related to the four groups do not represent absolute cutting points in this scale. The groups were established more on the basis of the clustering of products and on the distances or differences that appear between them.

Table 15. Weighted participation coefficients of main products in total value of agricultural circulation and computation of general priority index.

Product	Participation in total agricultural production value, 1972–76	Participation in			Weighted participation in			General priority index
		Domestic market (%)	Exports (%)	Imports (%)	Domestic market (%)	Exports (%)	Imports (%)	
Group 1								
Coffee	15.8	5.0	73.3	—	3.58	18.54	—	22.12
Beef cattle[a]	13.9	15.5	4.7	0.1	11.10	1.19	—	12.29
Dairy cattle	8.9	10.5	—	2.3	7.52	—	0.07	7.59
Group 2								
Cotton	4.9	5.1	6.2	—	3.65	1.57	—	5.22
Rice	5.9	6.8	0.8	—	4.87	0.20	—	5.07
Cassava	5.7	6.7	—	—	4.80	—	—	4.80
Pigs	5.1	6.0	—	—	4.30	—	—	4.30
Group 3								
Plantain	4.5	5.2	—	—	3.72	—	—	3.72
Sugarcane	3.2	3.1	5.8	—	2.22	1.47	—	3.69
Poultry meat	4.3	5.0	—	0.8	3.58	—	0.02	3.60
"Panela" (sugar loaf)	4.3	5.0	—	—	3.58	—	—	3.58
Eggs	4.0	4.7	—	—	3.36	—	—	3.36
Maize	3.6	4.2	—	3.5	3.01	—	0.11	3.12
Potato	3.5	4.0	—	—	2.86	—	—	2.86
Group 4								
Wheat	0.3	0.4	—	52.0	0.29	—	1.61	1.90
Bananas	1.9	1.2	2.8	—	0.86	0.71	—	1.57
Sorghum	1.3	1.5	—	—	1.07	—	—	1.07
Beans	1.2	1.2	0.6	0.2	0.86	0.15	—	1.01
Tobacco	0.9	0.7	1.9	—	0.50	0.48	—	0.98
Soybean	0.8	0.9	—	6.4[b]	0.64	—	0.20	0.84
Cocoa	0.7	0.8	—	6.9	0.57	—	0.21	0.78
African Palm	0.7	0.8	—	—	0.57	—	—	0.57
Barley	0.5	0.5	—	4.2	0.36	—	0.13	0.49
Yam	0.3	0.3	—	—	0.21	—	—	0.21
Sesame	0.2	0.3	—	—	0.21	—	—	0.21
Oats	—	—	—	2.4	—	—	0.07	0.07
Sheep	—	0.1	—	—	0.07	—	—	0.07
Relative importance of components of agricultural circulation		71.6	25.3	3.1				

[a] Includes live bovines.
[b] Includes soybean oil.
Source: Derived from Tables 12, 13, and 14.

cattle, and dairy cattle; group 2 (index value 4–7): cotton, rice, cassava, and swine production; group 3 (index value 2–4): plantain, sugarcane, poultry, "panela" (sugar loaf), eggs, maize, and potatoes; and group 4 (index value < 2): wheat, bananas, sorghum, beans, tobacco, soybean, cocoa, African palm, barley, yam, sesame, oats, sheep, and peanuts. Groups 1 and 2 are considered to be high priority for the country on the basis of their relative importance in the total value of agricultural circulation. Groups 3 and 4 represent medium and low priority products respectively.

Most of the products in groups 1–3 are food crops for direct consumption; cotton and sugarcane are used as raw materials for the manufacturing industry and coffee is intended primarily for export. Most of the foreign exchange produced by the export of agricultural products comes from crops in the first three priority groups.

Because the variables used in this model are basically production variables (i.e., production for the internal market, agricultural exports, and agricultural imports), two additional indicators were taken into account to see if they improved the analytical power of the model by substantially modifying the priority ranking established by the initial set of variables. The two additional variables considered were rural employment generated by each crop and the extension of land (area) under that crop's production.[22]

Table 16 compares the participation rates of the different crops in the total value of agricultural circulation (general priority index) with their relative importance in terms of the other two variables. Very few agricultural products undergo a change in their priority ranking important enough to warrant a reclassification in terms of general priority levels. As indicated in Table 16, only three products (plantain, maize, and "panela") shift from medium priority (group 3) to high priority (groups 1 and 2). Plantain and maize increase substantially in terms of both additional variables. The importance of "panela" (sugar loaf) is enhanced mainly by the employment it generates in the agricultural sector. The high ranking of maize in terms of the area under production should be interpreted with some reservation because the greater part of this area is shared with other crops (multiple cropping systems). Thus, the

net area actually used for maize would be much smaller.

The importance of tobacco increases to some extent in terms of employment generated (from low to medium), but it consistently ranks low in terms of the other two indicators. Thus, it would still remain as low priority.

The preceding analysis clearly shows that only minor modifications in the general priority ranking are introduced by the two additional variables. The overall ordering of products is maintained to a large extent.

Identification of Research Priorities within Selected Products or Problem Areas

The process for the identification of research priorities within selected products or problem areas was designed and carried out by the Instituto Colombiano Agropecuario (ICA) in 1979 and 1980. The first version of the National Plan for Agricultural Research (Plan Nacional de Investigación Agropecuaria) was published by ICA in January of 1981. A more detailed description and analysis of the methodology that was used in this process is presently being prepared by ICA.

Main Steps Followed in the Process of Identifying Research Priorities within Products: A Matrix Approach

As mentioned earlier, despite the fact that the formal identification of product or problem priorities had not been completed, the decision was taken in Colombia to go ahead with the determination of technological requirements and research needs at the product level.

In order to carry out this second phase of the planning process, the list of 28 agricultural products compiled by the Ministry of Agriculture (OPSA) was taken as a point of reference. Because these 28 products represent almost all of the agricultural production for which there is information, the proposed research programs cover a wide range of the present agricultural production.[23]

The process of identifying research priorities at the product level was carried out in four steps: (1) regionalization of the country into "ecologically homogeneous zones"; (2) characterization of each

---

[22] Rural employment generation is measured by multiplying the number of hectares under a given crop's production by the number of man-days of labour required per hectare. These are estimates published by the Ministry of Agriculture.

[23] The only major exceptions are coffee and sugarcane, which in the case of Colombia are research areas that are in the hands of the private sector.

Table 16. Comparison of the general priority index based on agricultural circulation with participation in area under agricultural production and employment generation.

Product	General priority index based on agricultural circulation	Participation in area under agricultural production	Participation in employment generated by agricultural sector
Group 1			
Coffee	22.12	26.6	17.2
Beef cattle	12.29	[a]	—
Dairy cattle	7.59	[a]	—
Group 2			
Cotton	5.22	7.0	7.3
Rice	5.07	9.6	5.9
Cassava	4.80	5.8	10.2
Pigs	4.30	—	—
Group 3			
Plantain	3.72	9.5	9.9
Sugarcane	3.69	2.1	2.9
Poultry	3.60	—	—
"Panela" (sugar loaf)	3.58	4.6	9.6
Eggs	3.36	—	—
Maize	3.12	15.6	12.2
Potatoes	2.86	3.3	6.4
Group 4			
Wheat	1.90	0.8	0.4
Bananas	1.57	0.5	1.5
Sorghum	1.07	4.7	0.8
Beans	1.01	2.8	2.2
Tobacco	0.98	0.8	7.1
Soybean	0.84	1.6	0.8
Cocoa	0.78	1.5	3.2
African Palm	0.57	0.5	0.9
Barley	0.49	1.7	0.3
Yam	0.21	0.3	0.7
Sesame	0.21	0.8	0.4
Oats	0.07	—	—
Sheep	0.07	—	—
Peanuts	—	0.1	—

[a] Livestock occupies approximately 25 million hectares, which implies that it would still remain in this high priority category in terms of the area under cattle production. Because it is so extensive in land use, this figure was not included for the determination of these percentages as it would drastically distort the overall picture.

Source: Balcazar, Alvaro and Torres, Ricardo. 1981. Selección de prioridades socio-económicas para la investigación agropecuaria. COLCIENCIAS.

region and analysis of the principal production systems that are found in them; (3) identification and analysis of the main "technological constraints" that have a negative impact on the production or productivity levels of the different products, under the specific environmental conditions that characterize each region; thus making the analysis both product-specific and region-specific; and (4) identification and analysis of potential research topics or issues that are considered to be important to solve the technological constraints faced by each product in specific regions.

The first three steps were carried out through a national survey, on the basis of which a technological profile or technological diagnosis of the agricultural sector was formulated.[24] The fourth step was carried out through working groups established for each product, in which the delphic technique was used (group discussions) for the identification and analysis of research topics or issues in response to the technological constraints previously identified.

In the first phase of the analysis, the country was divided into "natural regions" and "ecologically homogeneous zones," mainly on the basis of physical parameters that characterize and differentiate each zone. The principal physical parameters

---

[24] See ICA. 1980. Sector agropecuario Colombiano: diagnóstico tecnológico. Two volumes.

used in regionalizing the country were: climatic variables, water availability (hydrological resources), types of soil and soil characteristics, and dominant flora and fauna.

Seven main "natural regions" were identified within the country: Caribbean Region, Pacific Region, Andean Region, Inter-Andean Valleys, Orinoquia Region, Amazon Region, and Island Territories. Within each, an effort was made to identify subregions that could define ecologically homogeneous zones of economic importance (where relevant and only for the purpose of a more detailed analysis). These are geographical units that are more homogeneous from the point of view of the above-mentioned parameters.

The second and longest phase of this analysis was the characterization of these natural regions or ecologically homogeneous zones. This characterization covered several aspects:

(1) Characterization of the physical or environmental parameters, for example, climatic characteristics were analyzed in terms of: total and monthly precipitation levels (rain), temperature range and monthly variations, relative humidity, and sunshine. The soil characteristics were analyzed in terms of the dominant types of soil and in terms of such parameters as erosion, depth, external drainage, fertility (i.e., pH values), salinity, and elements that are low or in excess in the types of soil found in that region. The other aspects are characterized by similar parameters that are relevant for each case.

(2) Characterization of the socioeconomic characteristics of the region. Both economic and social aspects of the agricultural sector in the region were analyzed, such as: agricultural and animal production (both in terms of volume and in terms of its participation in national production); regional consumption and regional contribution to the national internal market and to exports; importance of agricultural production in the regional economy; economically active population, rural employment, and migration; land tenure structure and relationship with cropping and farming systems; and organizations of producers and managerial capacities.

(3) Characterization of the agricultural production system in that region. Identification and analysis of the principal agricultural products (both in terms of crops and animal production) and the principal farming systems and cropping systems that are being used. This leads to an analysis of the interaction between crops, cropping systems, and the environmental and socioeconomic characteristics of the region. Other aspects, such as the degree of mechanization, use of agricultural inputs, labour or capital intensity, productivity levels of the different crops or animals, energy sources, and forms and timing of planting and harvesting activities, are also taken into consideration in order to characterize the type of production technologies utilized.

(4) Characterization of the support services that exist in the region. This refers to such services as technical assistance, credit facilities, commercialization mechanisms, supply of agricultural inputs, transportation facilities, training institutions, and other support services.

The third step plays a central role in the process of identifying product research priorities because it is related to the identification and analysis of the main "technological constraints" that have a negative impact on the production or productivity levels of the different products under consideration. In order to do this, it was necessary to identify the principal technological factors that intervene in the production process, both in the case of crops and animal production.

In the case of crops, the principal technological factors were conceived in terms of eight categories, each one related to a specific discipline of the agronomical sciences. The eight technological factors are: (1) farming practices (including cropping systems); (2) production equipment: agricultural machinery and implements; (3) knowledge of plant genetics and the development of desirable genotypes and their seeds; (4) knowledge of insects, rodents, and molluscs, their impact on crops, and control methods; (5) knowledge of plant diseases, disease-causing agents (bacteria, virus, fungi), and control methods; (6) knowledge of plant physiology in order to improve their efficiency (yield) or control them (weeds); (7) soil as a factor of production; knowledge of soils: their characteristics, improvement, and conservation; and (8) water as a factor of production; knowledge of hydrological resources and water management and distribution (irrigation).

In the case of animal production, the following six technological factors were considered: (1) knowledge of animal production systems and techniques; (2) knowledge of animal physiology and reproduction; (3) knowledge of animal genetics and crossbreeding; (4) animal food and feeding systems; nutrition problems; (5) pasture and forage as a factor of production; (6) knowledge of animal diseases, their causes, and control.

In each region, an effort was made to identify and analyze the main "technological constraints" that have a negative impact on the production or productivity levels of the principal products (crops and animals) under the specific environmental conditions that characterize that region. These technological constraints were identified by analyzing the situation of each technological factor (either for crops or animals), as well as the impact of specific problems or bottlenecks identified in them on production or productivity levels. Thus, technological constraints were expressed in terms of limitations, deficiencies, or problems related to one of the technological factors that was responsible for low production or productivity levels (i.e., certain crops in a given ecological region or zone might be facing soil deficiency problems, or might show low yields or high vulnerability to diseases; or an important bottleneck for animal production in certain regions might be found to be poor pastures or inefficient animal production systems). These technological constraints lead to the identification of research needs and specific technological requirements (such as technical assistance) at the level of each product in given geographical regions (ecological zones) of the country.

These steps define an analytical matrix that permits different agricultural products to be related to specific technological constraints under certain environmental conditions that define ecologically homogeneous zones (Fig. 3). Each cell of the matrix in Fig. 3 defines a potential research area or topic, in order to solve a specific technological constraint (production problem) that is limiting the productivity level of a given agricultural product, within an identifiable region or ecological zone.

The same product may face different technological constraints, in different geographical or ecological regions. For example, in a given region the crop under consideration may face a serious problem of soil deficiency, whereas in other regions the main problem may be high vulnerability to diseases, despite relatively good soils. Furthermore, the importance of a given technological constraint may vary from one region to another for the same agricultural product. Thus, the analysis of technological constraints is both product-specific and region-specific, although some of them may cut across several regions.

It should also be noted that not all cells of the matrix are relevant, because not all products are found in all ecological regions or because a given technological constraint may not be relevant or important for all agricultural products (Fig. 3). The importance of each matrix cell (research topic) depends upon both the relative importance of the product and the magnitude (difficulty) and importance of the technological constraint to be solved.

The main output of the first three steps in the process of defining research priorities at the product level is the identification and description (diagnosis) of important technological constraints that limit production or productivity levels of specific agricultural products in certain ecological regions.[25] Further analysis of the importance of each research area (cell of the matrix), as well as the disaggregation of each area into more specific research topics (potential research projects), was carried out in the fourth step of the process.

Use of the Delphic Technique for the Identification of Technological Requirements and Research Needs

Having determined the principal technological constraints that limit the production or productivity levels of specific crops in certain ecological regions, the next step of the process was to establish research needs (and, therefore, research priorities). This implies a disaggregation of each of the matrix cells in Fig. 3 into research topics or projects that may contribute to the solution of each technological constraint.

To do this, special working groups were established in the different product and problem areas that were being considered. Each working group was made up of a group of experts with extensive experience in specific products and research areas, and with a good knowledge of the agricultural sector in the country and the production problems it faces.

These groups used the "Delphic" technique, which involved a group or panel discussion on the technological constraints under consideration, for the purpose of arriving at a consensus on the different aspects involved in each technological bottleneck and the research topics or projects that could contribute to the solution of those problems. This technique has been used widely in many countries, both in the identification of research needs and priorities, and in technological assessment (analysis of future technological developments and their impact).[26]

In this analysis, each group took into consideration the three major aspects that were identified in Fig. 1 as components of the general methodological framework for the identification of research priorities: (1) the technological constraints that have a negative impact on the production or productivity levels of specific agricultural products under the environmental conditions that characterize a given

---

[25] In the Colombian case this is presented in ICA. 1980. Sector agropecuario Colombiano: diagnóstico tecnológico. Two volumes.

[26] For a discussion of the use of the Delphi methodology and of matrix techniques in this type of analysis, see Cetron, M.J. and Bartocha, B. 1972. The methodology of technology assessment. New York, New York, Gordon and Breach Science Publishers.

Principal products considered to be of high socioeconomic importance or priority for the country	Principal technological constraints and ecologically homogeneous zones									
	Technological constraint 1			Technological constraint 2			Etc.	Technological constraint *i*		
	EHZ-1	EHZ-2	EHZ-*i*	EHZ-1	EHZ-2	EHZ-*i*		EHZ-1	EHZ-2	EHZ-*i*
Product 1										
Product 2					α / β					
Etc.										
Product *i*										

Fig. 3. Matrix approach to research planning in agricultural research. EHZ refers to the different ecologically homogeneous zones. α represents the importance of a given technological constraint, for a specific agricultural product, in a specified ecological zone or region. β represents the importance of the existing pool of knowledge and technological know-how that may be used in the solution of a specific technological constraint.

geographical region (demand for technology); (2) the pool of existing knowledge, know-how, and technologies (within the country or abroad) that is already available and that could be used to solve a specific technological constraint (supply of technology); (3) the desirable characteristics of technological change that one wishes to promote in the agricultural sector (desirable technological path); this provides criteria that may be used to evaluate technological alternatives, when they exist, or to design new technologies through research efforts.

The importance of the second factor is quite evident. In some cases, a technological constraint may be identified in a given product, despite the fact that there is technological know-how already available that could be used to solve the production problem under consideration. In such a case, the problem is one of transfer of technology to the producer and not of development of new technologies through research programs.

Each group, whose attention always centred on a specific product, had at its disposal three main inputs as a starting point for their deliberations:

(1) The technological diagnosis of the agricultural sector analyzes the production problems of the different crops, identifies major technological constraints, and makes a preliminary evaluation of the importance of each constraint.

(2) Brief state-of-the-art reports were prepared for each product (and, thus, for each group), sum-marizing the present research effort and the principal available technologies developed for that product. The objective of these reports was to have an approximate idea of the pool of existing know-how and technologies related to the product under consideration.

(3) The knowledge and experience each participant brought to the group. Given the importance of this factor, the selection of the group members is of crucial importance in this Delphi methodology.

The discussions of the working groups centred around two main issues: (1) analysis of the real importance and nature of each of the technological constraints that are confronted (each relevant matrix cell in Fig. 3) and (2) identification of research projects that should be carried out in order to generate the knowledge or know-how that is needed for the solution, elimination, or drastic reduction of that technological constraint.

With respect to the first issue, the importance and nature of the technological constraint under consideration was analyzed by comparing two indicators: the importance of the technological constraint that is being faced, from the point of view of its impact on production or productivity levels (α) and the importance or amount of the existing know-how that could be used effectively to solve or reduce the technological constraint (β).

The magnitude of these two indicators was "measured" in terms of an integral scale ranging in value from 1–10. In this scale, 1 represents a technological constraint of very low importance (impact), and a very low or limited technology supply. Ten represents a very important technological constraint (strong impact), and a highly important supply of technology that could be used to control or diminish the technological constraint under consideration. In both instances, 5 represents an intermediate situation. The values given to each technological constraint, with respect to these two indicators, were determined by each group on the basis of the three sources of information available to them. In terms of the analytical matrix presented in Fig. 3, every relevant matrix cell (each technological constraint identified) has these two values.

The range of points in both scales was divided into three categories: 1–3, low; 4–6, medium; 7–10, high. These three categories were used in the subsequent applications of the two indicators.

The comparison between the two indicators (α/β) in the case of each technological constraint was used to determine the importance or priority of that constraint, as well as some indication as to the nature of the technological problem faced. The different possible combinations of the comparison between the two indicators (α/β) were used to classify all identified technological constraints into three levels of priority (high, medium, and low), according to the relationship between the perceived importance of the technological constraint (α) and the present availability (supply) of know-how and technologies that could be used to control or diminish that constraint (β). The different possible combinations of this relationship (α/β) and their interpretation for assigning an overall level of priority to each technological constraint (matrix cell) are: high priority: medium/low, high/low, high/medium; medium priority: low/low, medium/medium, high/high; low priority: low/medium, low/high, medium/high.

An effort to formulate research needs and projects (the next step of the process) was carried out only for those technological constraints with high and medium priority levels. Low priority technological constraints were disregarded, except in those cases where a certain ongoing research level was considered necessary to maintain a technology previously developed.

In certain cases, an analysis of the relationship between α/β gives some insight into the nature of the technological problem that is being confronted. In the case of an important technological constraint, with a low availability or supply of technological know-how to cope with the problem, there is obviously a need for a research effort to develop the necessary technology. In the case of an important technological constraint (i.e., seriously limiting production or productivity levels) and the existence or availability of an important (high) or moderately important (medium) body of knowledge or usable technology to solve that constraint, the technological problem confronted is *not*, basically, a research problem (lack of knowledge).

In such a situation, the technologies that have been developed in the agricultural research stations (within the country or abroad) are not being used by the producers. Two major factors can explain this situation. Firstly, this may reflect a problem of inefficient agricultural extension and technology transfer to the producer. Thus, the technological requirement generated in this situation is not for more research but for better technology transfer mechanisms (technical assistance, credit, etc.).

Secondly, this situation may also be partly due to the fact that the technology that has been developed (existing supply) is not the most appropriate one for the type or characteristics of the producers for which it was developed. For example, the cost of the agricultural input (i.e., fertilizers) necessary to use that technology may be too high for the type of producer that should be using it, or the degree of mechanization or scale of production that are required do not correspond to the characteristics or capacity of the latter. In such a case, the conditions and characteristics of the producers themselves would require modification or alternative technologies more adapted to the production conditions existing in the country (research requirement) would have to be developed.

These two examples show that an analysis of the relationship α/β for each technological constraint may provide important insights into the nature of the technological problem confronted. Moreover, it also points out that not all technological requirements lead to research needs. They may also define problems with technological information and technical assistance or problems with diffusion and adoption of technologies.

The last step in this planning process was the identification and formulation of research topics or research projects that are considered important in order to control or diminish the production problem that is faced. As pointed out earlier, this last exercise was carried out only for those technological constraints that were considered to be of high or medium priority on the basis of the previous analysis. The research projects were identified and defined by each working group using the relevant information and inputs they had at their disposal. The group discussion technique and the expert advice provided by group members were used as a means for arriving at a consensus with respect to research projects (Delphi methodology).

Agricultural research[a]	Animal science research
Agricultural crops	*Animal species*
Sesame	Dairy beef cattle
Cotton	Specialized dairy cattle
Rice	Beef cattle
Peas	Poultry
Oats for forage	Swine
Cocoa	Sheep
Sugar loaf (panela)	Rabbit
Barley for malt	Bees
Barley for human feed	
Coconut	*Factors of production*
Cropping systems	Physiology and reproduction
Beans	Nutrition
Fruits	Animal production
Vegetables	Pasture and forage
Peanuts	Animal health
Maize	Animal genetics
Yam	Rural socioeconomic development
African Palm	Technology economic analysis
Potatoes	Socioeconomic factors determining the adoption of technology
Plantain	Production costs and retribution factors
Sorghum	Rural employment
Soybean	Formation and functioning of capital
Tobacco	Administration
Wheat	Demand and supply studies
Cassava	Product marketing
	Input marketing
Factors of production	Land size and tenure
Entomology	Types of guild organizations
Plant physiology	
Phytopathology	Rural communication
Plant breeding	Large producers
Soils	Private technical extension workers
Water and soil resources	Institutions related to formal and nonformal education in the rural sector
Farm processes	Change agents
Farm machinery	Small farmers

[a] This does not include two major research areas (coffee and sugarcane) because in the Colombian case these areas are in the hands of the private sector.

The outcome of this process was the formulation of a set of research projects for each agricultural product, aimed at solving or controlling the principal technological constraints for that product. The different research programs thus formulated constitute the National Agricultural Research Plan, recently presented in its first version.[27]

Some Observations with Respect to the National Agricultural Research Plan

Using this methodology, a first version of the National Agricultural Research Plan has been formulated in Colombia. The plan covers four main areas: agricultural research, animal science research, research on rural socioeconomic development, and research on rural communication. The two first areas are by far the most important components.

Each area is made up of a number of research programs, each one made up of a set of research projects. Not all research programs are formulated at the level of agricultural products. Some of them refer to the technological factors that were identified in the production of crops and animals, and to the agronomical disciplines that are related to them.

A total of 63 research programs (Table 17) were formulated with the following distribution in terms of the four areas previously mentioned: Agricultural research: 33 research programs, 25 of which involve

[27] See ICA. 1981. Plan nacional de investigación agropecuaria. Five volumes.

crops and 8 involving disciplines or factors of production. It should be noted that a research program on cropping systems was included as part of the 25 programs in terms of crops. Animal science research: 14 research programs, 8 of which deal with animal species and 6 with factors of production. Research on rural socioeconomic development: formal research programs were not formulated in this area but 11 research topics were identified as being of high priority for understanding rural socioeconomic development, and supporting technological development programs. Research on rural communication: 5 areas of research were identified dealing with the principal social actors or groups that intervene in the process of rural communication. The objective is to determine the characteristics and information needs of different types of users, relative efficiency of different communication media, and role of rural communication in the process of technology transfer.

The projects that are formulated within each research program are region-specific, in terms of the geographical regions into which the country was divided. For example, the 33 research programs of the agricultural area are made up of 638 research projects. These, in turn, are distributed among the different geographical regions as follows: Andean Region, 506 projects; Inter-Andean Valleys, 414 projects; Caribbean Region, 386 projects; Orinoquia, 125 projects; Pacific Region, 25 projects. A given research project can be related to two or more regions according to the distribution and importance of a crop or technological problem in the different regions of the country.

The large number of research programs and wide distribution of topics and research areas is one of the present problems or limitations of the first draft of the National Agricultural Research Plan. This is due to the fact that the first phase of the planning methodology described earlier has not been completed. The formulation of research programs at the product level (second phase) was carried out for almost all agricultural products and not only for those that are considered to be of high priority for the country.

Thus, although research priorities have been validly assigned within products or technological factors of production (second phase), this effort is still missing at the interproduct level, on the basis of socioeconomic priorities for research purposes (first phase). The consequence of this is the large number and wide distribution of research programs that characterize the present version of the research plan.

The last step of this planning process in the Colombian case will be the completion of the first phase of the methodology, using one or both of the analytical models discussed in this paper. This will presumably narrow down both the number and wide distribution of the research programs that will finally be included in the National Agricultural Research Plan.

Defining Research Priorities for Agriculture and Natural Resources in the Philippines

J.D. Drilon and Aida R. Librero[1]

Basic to the attainment of agricultural development is increased productivity both in terms of farm yield and the optimum utilization of available resources. A primary consideration is the promotion and support of research — coordinated, intensified, relevant, and applied to the needs of the country.

To be effective, a research program must respond to the current needs of developing agriculture and be sensitive to the needs of the future. However, research needs vary from one commodity to another. Obviously, the urgency of research undertakings differs between commodities, and variations in research needs also occur within a commodity. For example, a commodity might have been given greater emphasis in the past resulting in the development and adoption of better technology; therefore less research may be required at present. Another commodity might have lagged behind in terms of research and current needs might now call for greater research efforts on this commodity. This is especially true when new government programs and policies are implemented that necessitate more information and new technology for a particular commodity. This underscores the necessity of defining research priorities not only for more efficient research management and more effective and relevant research results but also for more meaningful allocation of limited resources, including research funds and manpower.

The definition of research priorities will be discussed in the context of the experience of the Philippine Council for Agriculture and Resources Research (PCARR). PCARR is the national agency

with a mandate from the Philippine government to plan and coordinate research in agriculture and natural resources. First, the tasks of PCARR and its organizational structure will be discussed in relation to defining priorities. The membership of the various bodies within the organization will be presented to provide the framework for understanding the role of scientists, academe, politicians, and the private sector in the process of defining research priorities. Second, the criteria for assigning priorities and the methods for making them operational are discussed. Third, the allocation of research funds and manpower is presented.

The Tasks of the Philippine Council for Agriculture and Resources Research

Up to the early 1970s an undesirable research situation existed in the Philippines that was characterized by: agricultural research that was not making a substantial impact on the economy despite the large sum of government funds that was being spent annually; uncoordinated activities with hardly any integrated planning among the various agencies; and fragmentary distribution and inefficient use of research resources.

This prompted the President to reorganize the national agricultural research system to make it a more effective tool for national development. Hence, the Philippine Council for Agricultural Research (PCAR) was established in November 1972 to provide for a systematic approach to the planning, coordination, direction, and conduct of the national research program in agriculture, forestry, and fisheries. Later, the functions of PCAR were expanded to include mines research resulting in the modification of its name to Philippine Council for Agriculture and Resources Research (PCARR).

---

[1] Director, Southeast Asian Regional Center for Graduate Study and Research in Agriculture (SEARCA), College, Laguna, Philippines, and Director, Socio-Economics Research Division, Philippines Council for Agriculture and Resources Research (PCARR), Los Baños, Laguna, Philippines, respectively.

Specifically, PCARR is entrusted with the following tasks: (1) define goals, purposes, and scope of research necessary to support progressive development in agriculture, forestry, fisheries, and mining for the nation on a continuing basis; (2) using the basic guidelines of relevance, excellence, and cooperation, develop the national agriculture and resources research program based on a multidisciplinary, interagency, and systems approach for the various component commodities; (3) establish a system of priorities for agriculture, forestry, fisheries, and mining research and provide meaningful mechanisms for updating these priorities; (4) develop and implement a fund-generating strategy for supporting agriculture and resources research; (5) program the allocation of all government revenues earmarked for agriculture and resources research to implement a dynamic national agriculture and resources research program; (6) provide the mechanism for assessment of progress and updating the national agriculture and resources research program; (7) establish, support, and manage the operation of a national network of centres of excellence for the various research programs in crops, livestock, forestry, fisheries, soil and water, mining resources, and socioeconomic research related to agriculture and natural resources; (8) establish a repository for research information in agriculture, forestry, fisheries, and mining; (9) develop a mechanism for full communication among workers in research, extension, and national development; (10) provide for a systematic program of agriculture and resources research, manpower development, and improvement; (11) provide for appropriate incentives to encourage top-notch research workers to remain working in their respective areas of agriculture and resources research; (12) enter into agreement or relationships with other similar institutions or organizations, both national and international, in the furtherance of the above purposes.

To effectively perform its designated tasks, PCARR was clothed with two vital powers: (1) the power to review all research proposals in agriculture and natural resources; and (2) the power to recommend research proposals to the Ministry of the Budget for funding. The second power was bolstered by a policy of the Ministry of the Budget that only research proposals recommended by PCARR would be eligible for government funding.

Complementing these powers is the mandate for PCARR to identify and coordinate the work programs of the network of research centres and stations throughout the country. To activate and strengthen the network, PCARR launched intensive infrastructure and manpower development programs that were aided in part by foreign institutions.

Organizational Structure of Research in the Philippines

To be able to carry out its functions effectively, the PCARR was organized into three main bodies: (1) the Governing Council (GC); (2) the Technical Program Planning and Review Board (TPPRB); and (3) the Secretariat.

Governing Council

The Governing Council (GC) formulates guidelines and policies for the national research program in agriculture and natural resources as well as for the operation of PCARR. It is composed of the following: Chairman — Chairman, National Science Development Board; Co-Vice-Chairmen — Minister, Ministry of Agriculture, and Minister, Ministry of Natural Resources; Members — Minister, Ministry of the Budget Representative, National Economic and Development Authority; President, Association of Colleges of Agriculture in the Philippines; Chancellor, University of the Philippines at Los Baños; Director General, PCARR; and two outstanding members of the private agricultural business sector.

Because of its composition, the Governing Council provides a stable link to the national science structure of the country. It assures responsiveness of PCARR to critical problems in agriculture and natural resources and provides for participation of virtually all sectors including education and the private sector. Above all, it provides for the participation of key government agencies and the private sector concerned with research and development to ensure relevance of policies to national development objectives.

Technical Program Planning and Review Board

The Technical Program Planning and Review Board (TPPRB) provides for the pooling of expertise and involvement of the technical sector, private sector, and top government policymakers in establishing a national program for research in agriculture and resources such as forestry, fisheries, and mines.

The TPPRB serves in an advisory capacity. Chaired by the PCARR Director General, it reviews the national research programs before they are referred to the GC for approval. The TPPRB directly links PCARR with the various agencies the members represent, or which are affected by PCARR's operations.

The TPPRB membership consists of: Director General, PCARR (Chairman); Executive Director,

National Food and Agriculture Council (NFAC) (Vice-Chairman for Agriculture); Director General, Natural Resources Management Center (Co-Vice-Chairman for Natural Resources); Deputy Director General for Research, PCARR (Ex-Officio Secretary); Director, Agricultural Programs Office, National Economic and Development Authority (NEDA) (Member); Chief, Planning and Service Division, National Science Development Board (NSDB) (Member); Chairman, Agriculture and Forestry Division, National Research Council of the Philippines (NRCP) (Member); Chief, Plans and Programs Services, Ministry of Natural Resources (NMR) (Member); Chief, Plans and Programs Services, Ministry of Agriculture (MA) (Member); Assistant Minister of Budget (Member); three research directors from different universities (Members); three representatives from the agricultural business sectors, preferably doing research work for industry planning (Members); two representatives of different small famers' groups (Members).

Agriculture and natural resources are appropriately represented in the Board.

Secretariat

The Secretariat of PCARR, which implements the policies and guidelines formulated by the Governing Council, consists of technical and nontechnical staff headed by a Director General, assisted by two Deputy Director Generals, seven research directors (crops, fisheries, forestry, livestock, mines, soil and water, and socioeconomics), and three service directors (applied communication, international projects, and administrative services).

The National Commodity Research Teams

A total of 34 commodity[2] research teams have been established.

Crops: (1) coconut; (2) corn and sorghum; (3) fibre crops; (4) fruit crops; (5) legumes; (6) ornamental horticulture and medicinal plants; (7) plantation crops; (8) rice and other cereal grains; (9) root crops; (10) sugarcane; (11) tobacco; and (12) vegetables.

Livestock: (1) beef-chevon; (2) carabeef; (3) dairy; (4) forage, pastures, and grasslands; (5) poultry; and (6) swine.

Forestry: (1) bamboo, rattan, fuelwood, and other nontimber products; (2) fibreboards and paper

products; (3) parks and wildlife; (4) reforestation and forest watersheds; and (5) timber products.

Fisheries: (1) marine fisheries; (2) inland fisheries; and (3) aquaculture.

Farm Resources and Systems: (1) water resources; (2) soil resources; (3) farming systems; and (4) agricultural engineering.

Socioeconomics: (1) applied rural sociology; and (2) macroeconomics.

Mines: (1) metallic minerals; and (2) nonmetallic minerals.

The PCARR makes effective use of available research talent in the country by drawing on outstanding scientists to participate on a short-term "on-call" basis for planning, coordination, review, and evaluation of the national research programs.

Each team has a commodity team leader who serves an average of 1 day per week monitoring the various commodity research programs. In addition, some 450 of the best available scientists in the country serve as members of the various teams for approximately 10 days per year. These teams take care of the basic planning, review, and updating of the various commodity research programs.

The Research Network

PCARR has a mandate to develop a national network of research centres and field stations to implement the national research program. PCARR has identified 128 of 500 research agencies and stations to comprise the national research network. These are classified into: (1) national research centres, which may either be single-commodity or multicommodity centres that conduct basic and applied research across a broad range of disciplines; (2) regional research centres, which conduct applied research for commodities of major importance in the region where the centre is located; and (3) cooperating field stations, which provide facilities and/or sites where adoptives trials or field experiments are undertaken to take into account microenvironmental differences.

At present, there are four multicommodity research centres, seven single-commodity research centres, eight regional research centres, and 130 cooperating agencies.

The members of the national research network include universities and colleges, agencies, and stations under the Ministries of Agriculture and Natural Resources and other institutions. These agencies are independent of PCARR except that their research projects are coordinated, monitored, and evaluated by PCARR.

---

[2] The term "commodity" is used very broadly to include physical products such as corn and fruits; resources like soils and water; and disciplines like rural sociology and macroeconomics.

Defining Priorities

Defining priorities for research is done at two levels, among commodities and within commodities. To be able to effectively assign priorities to the commodities they are classified as: (1) agriculture and natural resources; and (2) macrocommodities, which includes those that do not belong to (1) but nevertheless cut across several commodities under (1). These include soil resources, water resources, farming systems, agricultural engineering, applied rural sociology, and macroeconomics.

Criteria for Assigning Priorities to Commodities

Two sets of criteria have been identified, namely: (1) basic criteria that apply to both major commodity groupings; and (2) specific criteria that apply only to individual groups.

Basic Criteria

(1) Actual/potential contribution to sectoral value added. The greater the percentage contribution of the commodity to the gross value added from the sector the higher the point score for that commodity.

(2) Relevance to the socioeconomic programs of the government. This means that the data and/or model for improving the socioeconomic programs of the government are enhanced by the relevant findings derived from research on a particular commodity. The use of this criteria takes into consideration not only the number of government programs but also the scope and magnitude of the programs.

(3) Contribution to improved policy formulation and implementation.

(4) Links/support to other commodities. These links will be inclusive, i.e., backward, forward, and/or a combination of these.

(5) Contribution to employment. In general, the greater the employment generated by a commodity, the higher the score for that commodity.

(6) Contribution to improvement in labour productivity. Aside from the employment generated by a particular commodity, due consideration is given to improvement in labour productivity.

(7) Availability of research manpower and facilities. Manpower and research facilities must be available for the commodity. However, the availability or lack of research manpower and facilities must not hinder the inclusion/creation of a commodity but rather must serve as a reminder of the need for improving manpower in that area.

(8) Availability of appropriate technology. The less available appropriate technology there is, the higher the score for that commodity.

Specific Criteria

(1) Agriculture and natural resources: (a) contribution to export earnings; and (b) import substitution.

(2) Macrocommodities: (a) contribution to data base.

Each basic criterion carries a maximum of 10 points; each specific criterion carries 5 points of agriculture and natural resources and 10 points for macrocommodities.

A lot of statistical data is required to enable PCARR to assign priorities. The set of criteria includes both quantitative and qualitative variables. Data for some of the quantitative variables are available, e.g., value added, export earnings, etc. For some, projections have to be made based on available statistics from the findings of research projects.

For the qualitative variables, the individual member's (of the TPPRB) basic knowledge, inclination, and/or opinion is his primary guide in addition to the data, justifications, and projections provided by the technical divisions of PCARR. This in effect provides the sociopolitical input in the decision-making process.

Process for Assigning Priorities

The PCARR secretariat makes the preliminary steps. Then the individual members of the TPPRB make their own scoring of the various commodities. The process, therefore, incorporates the inputs of the TPPRB membership, which includes scientists, academe, administrators, and the private sector.

From the TPPRB, commodities that are given priority are submitted to the governing council. The 34 commodities are classified into three groups: (1) priority I; (2) priority II; and (3) priority III. Because rural sociology and macroeconomics encompass all other commodities and are considered to be important in agriculture and natural resources development, they were classified as priority I but given a special group status. Hence, we have socioeconomics and special projects. It was recognized that special projects and urgent research may be needed at any time because of such things as: the discovery of a potentially important nontraditional commodity; new development programs to be implemented by the government; or unexpected natural catastrophes like typhoons or diseases. Some allowance was made for such projects.

As expected, when the commodity classification was submitted to the governing council, a movement from one group to another was made depending on the programs, interests, and other justifications of the members of the council. This is part of the "political input" in the process. In fact, even at the

TPPRB level, some political considerations have been taken into account.

Commodity Priorities

The 34 commodities are classified below according to the groupings made, that is: priority I; priority II; priority III; and socioeconomics and special projects.

Priority I: (1) coconut; (2) corn and sorghum; (3) fibre crops; (4) legumes; (5) root crops; (6) sugarcane; (7) vegetables; (8) aquaculture; (9) marine fisheries; (10) forage, pasture, and grasslands; (11) cara beef; (12) nontimber products; (13) reforestation and forest watershed; (14) timber products; (15) metallic minerals; and (16) agricultural engineering.

Priority II: (1) fruit crops; (2) rice; (3) tobacco; (4) beef chevon; (5) inland water; (6) parks and wildlife; (7) farming systems; (8) soil resources; (9) water resources; and (10) nonmetallic minerals.

Priority III: (1) ornamental horticulture; (2) plantation crops; (3) dairy; (4) pork; (5) poultry; and (6) pulpwood, fibreboards, and paper products.

Socioeconomics and special projects: (1) applied rural sociology; and (2) macroeconomics.

Resource Allocation

With these priorities, the PCARR is able to allocate the limited funds available for research among the different commodities. As a guideline, the council has allocated 80% of the annual total research budget to priority I commodities, 10% to priority II, 3% to priority III, and 7% to socioeconomics and special projects.

Priorities Within Commodities

Because resources are limited and many problem areas must be studied for each commodity, PCARR defines the national commodity research program and identifies the priority research areas. Priorities are based on three major factors: (1) objectives of the national development plan; (2) status of knowledge/technology; (3) requirements of national development programs.

Among others, the objectives of the national development plan include self-sufficiency in food, increased income, improved income distribution, better employment opportunities, and improved nutrition. Such objectives are basic to the determination of priority research areas.

The state-of-the-art in a particular commodity defines what knowledge or what technology are presently available or adopted by the end-users and what gaps exist in the knowledge. End-users include producers, policymakers, and others. In general, the primary target of the research program is the small producer.

Various development programs are being implemented by the government. These programs have to be evaluated to provide guidelines for modification and improvement. Furthermore, problems must be identified so that solutions can be provided.

Refining the research program for each commodity is a major responsibility of the commodity research teams. These teams are multidisciplinary and multiagency in composition and thus provide a systems approach to the process.

Research priorities are further discussed with regional officials of the Ministries of Agriculture and Natural Resources and the National Economic and Development Authority to assure relevance and comprehensiveness of the research program.

All commodity research priorities are reviewed by the TPPRB before submission to the Governing Council. These priorities are updated every year to account for new developments and for new programs of the government.

Manpower Resources

Because of the need to develop scientific research manpower to effectively implement the national agriculture and natural resources research program, studies on the available manpower resources in agriculture, fisheries, forestry, and mines research have been conducted since 1971. The total available manpower resources increased from 2234 in 1974 to 3046 in 1978, an increase of about 36% in 4 years.

The distribution of researchers has remained virtually the same through the years, with the greatest number (1431 or 47% in 1978) found in colleges and universities. The Ministry (then Department) of Agriculture employed 24% and 17% in 1974 and 1978, respectively; whereas the Ministry of Natural Resources had 7% in 1974 and 13% in 1978. In 1978, the remainder were distributed among the National Science Development Board (4.5% in 1978), other government agencies (10%), international agencies (8%), and private agencies (0.49%) (Table 1).

The majority of the researchers hold a Bachelor of Science degree (67 and 72%, respectively, in 1974 and 1978) and in 1978 only 8% had a doctorate, which was a decline from 11% in 1974.

Researchers with specialization in crop sciences increased from 608 in 1974 to 1045 in 1978 and comprised the biggest group (31% of the total). Researchers with specialization in food and nutrition remained the smallest group both in 1974 and 1978 (3% and 2%, respectively). However, social science researchers decreased from 31% in 1974 to only 17% in 1978 (Table 2).

The current research manpower spends about half of its time doing actual research. The actual

Table 1. Distribution of researchers by agency and degree of training, 1974 and 1978.[a]

	B.S.		M.S./M.A.		Ph.D.		Total	
	1974	1978	1974	1978	1974	1978	1974	1978
Colleges and Universities	488	811	397	428	213	192	1098	1431
Department of Agriculture	491	484	49	39	4	4	544	527
Department of Natural Resources	153	346	8	21	—	1	161	368
National Science Development Board	135	113	22	20	14	4	171	137
Other government agencies	76	277	24	18	2	4	102	299
Internatinal agencies	121	165	4	64	1	40	126	269
Privates agencies	24	11	5	3	3	1	32	15
Total	1488	2207	509	593	237	246	2234	3046

[a] Source: Manpower Resources in Philippine Agriculture, Fisheries, Forestry, and Mines Research, 1974 and 1978.

utilization of research manpower was measured in terms of scientist-man-year (SMY), which is the percentage of full-time equivalent devoted to actual research work. Hence, the 3046 researchers spent only roughly 50% of their time on research.

In 1974, researchers in universities and colleges registered the lowest SMY (34%). This could be explained by the inherent complimentarity of teaching, extension-education, and research in these institutions that results in less time being spent on research by faculty members. Likewise, some faculty members, especially those with advanced academic training, have been drafted to undertake part-time assignments in government planning bodies and action agencies and this further reduces the time available for research.

Manpower Development Program

Development is people-centred. People are the target of development as well as its implementer and facilitator. Together with the material inputs, human resources are part and parcel of all development activities and concerns. The success of development efforts hinges on the provision and harnessing of an adequate number of people with appropriate skills, training, knowledge, and orientation. This requires a concerted, directed effort to continually educate and train the people who conduct and support development-related activities. PCARR therefore pursues a manpower development program for agriculture and resources research personnel in support of current national development thrusts.

Started in 1973, PCARR's manpower program is aimed at developing the agricultural and resources research capability of the Philippines. The goal is the effective implementation of a national agricultural and resources research program in the country through upgraded and improved research manpower, facilities, and streamlined research management and operations.

The program is of two kinds: (1) degree (includes courses at the B.S., M.S./M.A., and Ph.D. levels); and (2) nondegree, which includes training courses, study tours, observation trips, and participation in conferences, symposia, or similar development-oriented courses.

Because of limited financial resources, the primary target of the program is the identified national research network, including the PCARR Secretariat. However, since the national research system is not confined to this network, a modest percentage of the total resource is set aside for nonmember agencies and the private sector.

Table 2. Distribution of researchers by area of specialization, 1974 and 1978.[a]

	1974		1978	
	Number	Per-cent	Number	Per-cent
Crop sciences	608	27	1045	31
Social sciences	681	31	582	17
Forestry sciences	223	10	446	13
Animal sciences	220	10	227	7
Soil and water sciences	164	7	393	12
Fishery and oceanic sciences	139	6	296	9
Physical and chemical sciences	123	6	106	3
Food and nutrition sciences	76	3	69	2
Biological sciences	—	—	197	6
Total	2234	100	3361[b]	100

[a] Source: Manpower Resources in Philippine Agriculture, Fisheries, Forestry, and Mines Research, 1974 and 1978.
[b] More than the total number of researchers surveyed because of overlapping of field of specialization.

Table 3. Number of scholars by field of specialization (1973–74 to 1980–81).

	Ph.D.	M.S.	B.S.	Total
Agriculture				
Crop production	15	53	—	68
Crop protection	13	45	—	58
Crop improvement	2	26	—	28
Soil science	7	33	—	40
Agri. engineering	—	33	—	33
Agri. education	10	10	—	20
Rural sociology	4	11	—	15
Animal science	8	22	—	30
Plant physiology/ Botany	7	17	—	24
Agricultural economics	2	22	—	24
Dev. communication	5	12	—	17
Agribusiness/Business administration	2	14	—	16
Food science	—	7	—	7
Statistics	—	13	—	13
Human ecology	—	4	—	4
Agricultural chemistry	1	5	—	6
Seed technology	—	2	—	2
Agrarian studies	—	2	—	2
Management/ Administration	—	1	—	1
Forestry				
Forest resources management	9	32	—	41
Forest biological science	1	15	—	16
Wood science and technology	—	8	—	8
Wood chemistry	—	1	—	1
Wildlife chemistry	—	2	—	2
Fisheries				
Aquaculture	—	52	2	54
Inland fisheries	—	—	43	43
Marine biology	—	33	1	34
Food science/Chemistry	—	10	—	10
Oceanography	—	1	—	1
Zoology	—	1	—	1
Total	86	487	46	619

The awarding of scholarships for degree programs is based primarily on the commodity assignment and research program of the respective research stations. In addition, applicants for scholarships are evaluated on: (1) educational qualifications, (2) professional/work experiences; and (3) potential and future plans.

Progress of manpower development program as of 1980–81

From a start with nine scholars from the Southern and Central Luzon Regions in the second semester of 1973–74, the degree program has expanded to other regions. A total of 619 awards have been made as of the first semester of 1980–81. Of this number, 243 (39%) were from Southern Luzon; 28 (5%) from Central Luzon; 120 (19%) from Visayas; 100 (16%) from Mindanao; and 40 (7%) were unattached.

Table 3 lists the fields of study pursued by PCARR scholars. Crop sciences scholars comprised the biggest group, 192 (31%). Those with specialization in fishery and oceanic sciences numbered 143 (23%); social sciences 100 (16%); forestry sciences 68 (11%); animal sciences 30 (5%); and soil and water sciences 67 (10.82%). A mere 19 (3%) of the total number of scholars specialized in physical and chemical sciences.

Of the total number of degree-oriented local awards made by PCARR, almost all the scholars came from the National Research Network with cooperating stations having 241 (39%) and multicommodity agencies 135 (22%). Other government agencies had 9 (1.5%); unattached 40 (6%); and the PCARR Secretariat 32 (5%).

For short-term training PCARR either finances the participation of researchers and support staff for training abroad or sponsors the training itself locally. The 72 foreign nondegree training awardees were distributed as: 40 for short-term programs; 19 for conferences/seminars, and 13 for observation/study tours. These awards vary in topic and clientele level. Among others, topics included natural resources, economics and policy, management, education and human resource development; and production and technology. Administrators, researchers, and support staff used the program.

Agricultural Research Resource Allocation Priorities: The Nigerian Experience

F.S. Idachaba[1]

This paper examines selected features of research resource allocation priorities in a national agricultural research system, using Nigeria as a case study. Specifically, the paper (1) examines the allocation of resources to national agricultural research within a macroeconomic context; (2) specifies criteria for determining research resource allocation priorities within a national agricultural research system, given the usual budget constraint of the funding agency; (3) specifies criteria for assigning agricultural research priorities among different tiers of government, especially within a federal framework; and (4) examines the much neglected issue of research priority lags.

Food-deficit countries are, in their efforts to accelerate food production, increasingly running into agricultural research bottlenecks. Several food production, processing, and distribution bottlenecks are not reflected in current agricultural research resources allocation priorities. Resource allocation priorities lag behind the emerging constraints of the national food systems of many food-deficit countries, and many allocative puzzles remain even where the political leadership and national research management wish allocative priorities to reflect the new emerging constraints. The specification of criteria for determining research priorities therefore assumes new policy and professional relevance.

Very little exists in the literature on criteria for assigning agricultural research priorities among different tiers of government. This paper provides an explorative specification of some criteria, derived from economic theory, to guide the allocation of agricultural research responsibilities among different tiers of government.

Nigeria has a long history of agricultural research (dating back to 1893), is heterogeneous, and offers many emerging food problems and challenges to its national agricultural research system. The lessons from the Nigerian experience should have significance for other smaller food-deficit countries.

An Analytical Framework

Unless we are guided by some analytical framework, we may be victims of "empirical *ad hockery*," an unstructured effort, on a country by country basis, to try to understand the research resource allocation process.

Research Priorities Determined by Policy Objectives

It is postulated that the development process is guided by systems of policy objectives. At the highest level of aggregation are the basic goals of the society: its political values, a statement of the desired form of government, perhaps including a statement of ideology and also possibly a statement about preferred patterns of distribution of wealth and income among the nation's citizenry, etc. From the societal goals usually flow a set of macroeconomic development objectives: the former get transformed into the latter. In general, the macroeconomic objectives get transformed into sectoral development objectives, including agricultural sector objectives. These agricultural sector objectives get transformed into subsectoral objectives that in turn get transformed into program objectives, including agricultural research objectives. Specification of agricultural research objectives determines agricultural research priorities. This sequential process of transformations of policy objectives enables us to link "lower level" objectives (e.g., agricultural research) with "higher level" objectives (e.g., societal goals).

This way of looking at the process of specifying agricultural research priorities has several advantages. First, it ensures consistency between

[1] Department of Agricultural Economics, University of Ibadan, Ibadan, Nigeria.

"lower level" priorities and "higher level" priorities. This streamlines the resource allocation process from the highest level of aggregation to the lowest level — the project and subproject levels. It minimizes arbitrariness in actual research resource allocations by the political leadership or the research management itself. Second, this transformation process enables changes in "higher level" objectives to be translated or transformed into "lower level" research priorities. For example, a societal goal of self-reliance could get translated into increased allocations to agricultural research that would make the country less dependent on other countries for some key food imports such as wheat and long-grain rice. Third, it enables us to predict the implications of changes in "higher level" objectives for "lower level" priorities. It enables us to predict needed adjustments in national agricultural research management in response to changes in societal, macro-economic, and agricultural sector goals.

A few examples will illustrate the transformation process. A macroeconomic objective of rapidly increasing per capita income transforms into a rapid increase in per capita agricultural production. Accelerated increases in per capita agricultural production require: (1) supportive developmental research aimed at generating new high-yielding seeds and livestock breeds together with complementary work in agronomy, crop physiology, crop nutrition, and animal husbandry; (2) maintenance research to prevent crop and livestock losses from pests, diseases, and environmental stress;[2] and (3) institutional research into supporting institutional infrastructures both in public research management accounting and in rural institutions.[3]

A macroeconomic objective of raising nutritional levels translates into research into efficient sources of calories and proteins. It is not sufficient that per capita available calorie supplies be adequate, on the average. In addition, policy aims at minimizing uncontrollable and undesirable fluctuations in nutritional levels. This again gets transformed into agricultural research priorities that not only seek efficient sources of calories but sources that are not highly susceptible to environmental stress, particularly moisture.

For sociopolitical reasons, developing countries may not wish to have major regions of their countries left behind in the development process — the desire to minimize wide disparities in regional developments. One transformation of this objective is massive investments in irrigation and water resources development in drought-prone areas. This transforms into new research emphasis on irrigation agriculture, problems of large-scale river basin development, etc.

Finally, a country may declare "self-reliance" as a major societal goal within the context of economic nationalism. One concrete transformation of this goal is self-sufficiency in basic staples. This transforms into new research priorities on all aspects of production and processing of those foods of which the country is a major importer.[4]

Transformation of Constraints

Agricultural research priorities are determined through a sequential transformation of policy objectives. However, the policy process in food-deficit countries is one of constrained optimization of policy objectives. Agricultural research priorities get determined by the constraints of the policy optimization process. Given the sequential set of policy objectives running from the "highest level" (societal) to the "lowest level," three main classes of constraints can be identified: production constraints; distribution constraints; and institutional constraints.

Production Constraints

With the determination of commodity research priorities, the next step is the identification of input research priorities aimed at removing those constraints centring on resource productivity: technical determinants of productivity and yield coefficients, including maintenance and loss prevention, and the organizational determinants of resource productivity that centre on the determinants of allocative efficiency. Research priorities to tackle these constraints include the structure, conduct, and performance of farm labour markets and their interactions with institutionalized and well-organized urban labour markets and the structure of rural capital markets, including accessibility of credit facilities to the majority of farming households and farming systems research, etc.

---

[2] For the development/maintenance research classification, see Evenson, R.E. and Kislev, Y. 1975. Agricultural research and productivity. New Haven, Connecticut, Yale University Press.

[3] These are not water-tight classifications because agricultural research is essentially multidisciplinary: it serves little purpose for instance for the rice breeder to come up with a new high-yielding rice variety that is susceptible to rice-blast disease.

[4] See Idachaba, F.S. 1980. Agricultural research policy in Nigeria. Washington, D.C., International Food Policy Research Institute.

Distribution Constraints

Research priorities centre on identification and prevention of main distribution losses, processing bottlenecks related to technical inefficiencies, and sexual role differentiation.

Institutional Constraints

Research priorities centre on specification of the most effective institutional arrangements for implementing identified research priorities, the role of inhibiting and facilitating rural institutions, etc.

Assigning Responsibility for Agricultural Research

In the real world, different levels of government provide different levels of resource support for a national agricultural research system. Rational guidelines are required for assigning executive research responsibilities among state and federal governments. Schultz (1971) defined agricultural research "as a specialized activity requiring special skills and facilities that are employed to discover and develop special forms of new information, a part of which acquires the properties of economic information."[5] Research can also be distinguished as: basic; applied; and adaptive.

Basic research yields information mainly in the form of new ideas, concepts, and models, the benefits from which are usually not specific enough to be appropriated by the researcher or the funding agency. Examples include biological nitrogen fixation research, general agrometeorology studies, genetic manipulation, physiology of nutrient absorption, and biology of pests. In these types of research, it is virtually impossible to exclude "free riders": firms, individuals, and residents of other states in a federation from appropriating the benefits of such basic research.

Applied research, on the other hand, yields results that are specific enough to be appropriated by the researcher or the funding agency — "free riders" can plausibly be excluded. Examples include varietal selections and trials in selected environments and insecticidal trials. A prospective funding agency, such as a state government, can relate expected appropriable applied research findings to re-

search resources that have opportunity costs to residents of the state.[6]

Assume a federal setup with component states in which state governments, democratically elected, seek to maximize the social welfare of citizens of their state. Residents of a state are presumed to be concerned with the allocation of limited agricultural resources that are of state (practical) relevance — and that they would therefore fund research projects whose expected benefits can be seen to be appropriable by citizens of the state.

Funding of basic research by a state yields research results that are appropriable by other beneficiaries in other states of the federation. It is not feasible — if only because of prohibitive transactions costs — to exclude "free riders." There will therefore be underinvestment in basic agricultural research if left to individual states.

This reasoning leads to the following proposition: the more ecologically diverse a country is, the more basic agricultural research should be supported by the national (federal) government, leaving applied and adaptive research to lower cells of government.

The collective demands by residents of a state within a federation for the fruits of basic research can be represented by the demand curve DD in Fig. 1. This demand curve reflects the social valuation of research benefits by the state's residents. Let SS represent the supply curve of basic agricultural research — the more basic research that is already being provided, the more expensive it becomes to produce additional basic research output, either because it becomes increasingly more expensive to hire additional good scientists at the margin or, as is so common in developing countries, because basic research rapidly runs into diminishing returns because of limited research management capability or a limited number of technical support staff, laboratory technologists, and supplies. The optimum amount of basic agricultural research that will be supported and financed by a state government is given by OM.

The demand for basic research by the whole society, including residents of other states within the federation, is represented by DD[1] — reflecting the fact that benefits to the whole federation from basic

---

[5] Schultz, T.W. 1971. Efficient allocation of resources in agricultural research. In Fishell, A., ed., Allocation of Resources in Agricultural Research. Minneapolis, Minnesota, University of Minnesota Press.

---

[6] Here again, the distinctions are not water-tight compartments. For example, fertilizer trials in lowland rice in Nigeria have revealed that nitrogen in sulfate of ammonia was rapidly leached with the onset of rains and floods. Researchers then backed up to investigate the processes of slowing down nitrogen release so plants could utilize the nutrients during their critical periods. Sulfur-coated urea is being tried as a substitute for sulfate of ammonia.

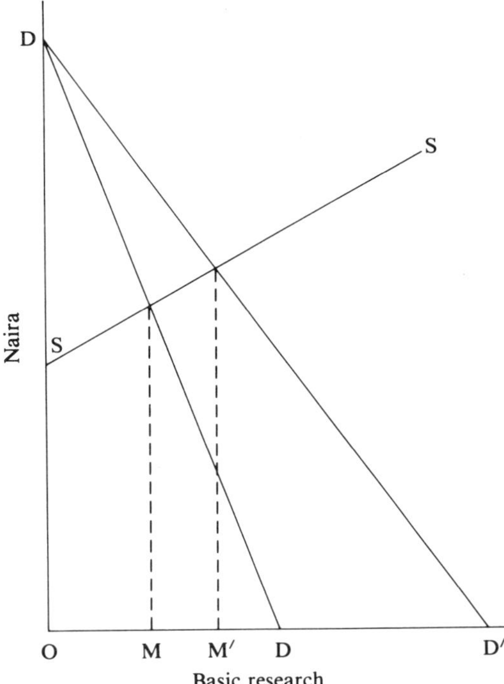

Fig. 1. Demand and supply of basic research in a federal setup.

research are greater than the benefits accruing to the residents of a given state that finances such research.[7] From the country's viewpoint, the optimum amount of basic research is OM^1, where the marginal social value of basic research equals the marginal social cost. This model is predicated on the plausible assumption that a state government does not allocate resources to basic research not caring who reaps the benefits: resources are presumed to be committed with the needs and welfare of residents of the state in mind.

The converse of the proposition holds: the more narrowly confined (restricted) the production of a crop in terms of its agroclimatological and agrobotanical characteristics, the more basic (as well as applied and adaptive) research into the crop should be conducted by lower cells of government, e.g., state governments.

The Nigerian experience with maize and rice research has been consistent with the original prop-

osition. Most of the research on maize and rice, crops that are grown in all the ecological zones of the country, has been supported by the federal government at Moor Plantation and Badeggi. The Northern and Western Regions supported some applied research on maize and rice at the Institute for Agricultural Research (IAR) and the Institute of Agricultural Research and Training (IAR&T). Research on groundnuts and cotton, the production of which is concentrated in the North, was until 1975, the financial responsibility of the Northern Regional Government. The old Western Regional Government was responsible for much of the research support on cocoa, a crop confined mainly to Western Nigeria.

Agricultural Research Priorities in Nigeria

A botanical research station was established in Lagos in 1893. By 1899, a model farm was established at Moor Plantation to propagate rubber trees and general agriculture. The Nigerian agricultural research system has grown from this humble beginning to the present situation where there are now 18 agricultural research institutes under the Federal Ministry of Science and Technology (Table 1).

Balance Between Crops, Livestock, Fisheries, and Forestry Research

For most of Nigeria's agricultural research history, agricultural research priorities could not be derived from any articulated set of economic development objectives or basic goals of the society. Any transformation process was only implicit (see Tables 2 and 3). Prior to 1965, there was no federal institution with responsibility for agriculture; there was no national agricultural policy. The little coordination that existed was in the area of agricultural research and even that was done in the Federal Ministry of Economic Development. Agricultural research priorities could not be said to be sequentially derived from "higher level" objectives as in the analytical model: allocations for agricultural research were to service the enclave export economies of the Western, Eastern, and Northern Regions.

Evaluation of Relative Allocations of Research Resources to Subsectors

Two criteria are used for determining research priorities among crops, livestock, fisheries, and forestry: past and projected roles of subsectors in the national economy, and the place of the subsector within the specific framework of national nutritional and food policy.

---

[7] For a similar approach in the case of the individual and the society, see Sjaastad, L. The economics of basic and applied research. Unpublished paper, University of Chicago.

Table 1. The distribution of agricultural research institutes, Nigeria, 1981.

Food Crops

Institute for Agricultural Research (IAR), Samaru, Kaduna State

National Cereals Research Institute (NCRI), Ibadan, Oyo State

National Root Crops Research Institute (NRCRI), Umudike, Imo State

National Institute for Horticultural Research (NIHORT), Idi-Ishin, Imo State

Institute for Agricultural Research and Training (IAR&T), Ibadan, Oyo State

Tree Crops

Cocoa Research Institute of Nigeria (CRIN), Gambari, Oyo State

Nigerian Institute for Oil Palm Research (NIFOR), Benin, Bendel State

Rubber Research Institute of Nigeria (RRIN), Iyanomo, Bendel State

Forestry Research Institute of Nigeria (FRIN), Ibadan, Oyo State

Livestock

National Veterinary Research Institute (NVRI), Vom, Plateau State

National Animal Production Research Institute (NAPRI), Shika, Kaduna State

Nigerian Institute for Trypanosomiasis Research (NITR), Kaduna, Kaduna State

Leather Research Institute of Nigeria (LRIN), Zaria, Kaduna State

Fisheries

Lake Chad Research Institute (LCRI), Maiduguri, Borno State

Kainji Lake Research Institute (KLRI), New Bussa, Kwara State

Nigerian Institute for Oceanography and Marine Research (NIOMR), Lagos, Federal Territory

General Services

Agricultural Extension and Research Liaison Services (AERLS), Samaru, Kaduna State

Nigerian Stored Products Research Institute (NSPRI), Lagos, Federal Territory

Relative Importance in the National Economy

From Table 3, crop research should have had top priority. Research on agriculture (crops), which averaged 49.88% of GDP in the 4 years preceding the launching of the 1962–68 Plan and 45.58% during 1962–68, was allocated only 12.76% of federal expenditures on crops; the crop research allocation as a percentage of total federal public sector budget during the Plan was only 1.62%. Thus in terms of both the historical and plan period importance of agriculture (crops) in the national economy, federal allocations for crop research in the 1962–68 Plan were grossly inadequate. Allocation to crop research in the 1970–74 Plan was also inconsistent with the relative importance of the crops subsector in the years preceding the plan period and during the plan period itself. Agriculture averaged 39.96% of GDP at current factor costs during 1968/69–1969/70 and 34.70% during 1970/71–1972/73. Its allocated share of total federal expenditures on all sectors during 1970–74 was only 0.86%. Crops were projected to contribute 20.78% of GDP during 1975–80, but crop research was only allocated 0.16% of all federal expenditures during the Plan. Thus, during the last two decades when structural forces required an increase in the allocations to agricultural research relative to other items of federal government expenditure on crop production, to counter the declining growth rate of the agricultural sector, the nation witnessed a drastic reduction in the relative allocations to crop research.

The pre-1962 period witnessed more emphasis on veterinary, fisheries, and forestry research than was warranted by their relative contributions to the GDP. Livestock research as a proportion of federal expenditures on livestock fell from 26.10% in 1953/54 to 9.23% in the 1975–80 Plan period. Planned livestock research expenditures in 1962–68 were 100% of all planned expenditure on livestock by the federal government. The fall in the percentage allocation to livestock research reflects the more than proportionate rise in federal expenditures on other nonresearch federal programs. This means that the federal government planned no direct activity expenditures in livestock during the period.[8] In fisheries, the share of fisheries research in federal expenditures on fisheries fell from 59.26% in 1953/54 to 18.79% in the 1975–80 Plan. In forestry, forestry research was 7.18% and 12.51% of federal expenditures on forestry in 1953/54 and in the 1975–80 Plan, respectively.

When all research categories are aggregated, allocations for research on agriculture (crops), livestock, fisheries, and forestry fell from 20.97% of all federal expenditures on these subsectors in 1953/54 to only 5.74% in the 1975–80 Plan. These historical

---

[8] This unusually high percentage may reflect failure of the plan document to differentiate between research and nonresearch expenditures in livestock.

Table 2. Relative shares of federal allocations for agricultural research in Nigeria.[a]

	1953/54	1954/55	1955/56–1959/60	1962–68 Plan	1970–74 Plan	1975–80 Plan	1975–80 Plan/Revised
Crop research (as % total research)	37.42	39.02	63.08	49.04	63.07	60.89	57.34
Veterinary research (as % total research)	39.40	37.80	20.36	10.79	16.71	25.70	27.29
Forestry research (as % total research)	12.58	11.59	11.68	16.69	12.61	3.79	5.02
Fishery research (as % total research)	10.60	11.59	4.87	23.47	7.62	9.61	11.38
Crop research (as % total fed. exp. on crops)	28.18	24.85	77.14	12.76	11.23	6.54	4.24
Vet. research (as % fed. exp. on vet. and livestock)	26.10	22.75	16.28	100	75.66	11.96	9.23
For. research (as % fed. exp. on forestry)	7.18	6.85	6.99	NA	36.07	10.18	12.51
Fish. research (as % fed. exp. on fisheries)	59.26	61.29	23.36	56.27	84.70	13.23	18.79
Total research (as % fed. exp. on crops, for., and livestock)	20.97	19.56	25.33	22.66	16.13	7.96	5.74
Total research (as % fed. exp. on all public sectors)	—	—	—	3.30	0.99	0.31	0.28

[a] Source: Idachaba, F.S. 1980. Agricultural research policy in Nigeria. International Food Policy Research Institute, Washington, D.C.

allocations are clearly inconsistent with the dominant position of agriculture, livestock, fisheries, and forestry during the period. They are also inconsistent with a developmental strategy that relies on millions of small-scale farmers for the nation's agricultural output. The average annual relative contribution of this sector to GDP at current factor cost in the years before the launching of all the development plans was: 63.49% during 1958/59–61/62; 53.42% during 1965/66–69/70; and 28.8% during 1970/71–74/75. Yet, in spite of this evidence to successive generations of authors of the development plans, research allocations to the agricultural sector accounted for only 3.3, 0.99, and 0.28% of planned federal expenditures in all sectors in the 1962–68 Plan, 1970–74 Plan, and 1975–80 Plan, respectively.

For every naira projected by the plan as being generated by the agricultural sector in the 1975–80 Plan, only 0.0042 of one naira[9] (or 0.42 of one kobo) was being allocated for agricultural research! To note that this is grossly inadequate is an understatement.

Nutritional Importance of Subsectors

The earliest available estimate of per capita calorie intake of the Nigerian population was for 1952/53: 2250 calories per capita per day. Protein availability per capita in the same year was estimated at 50 g per day, made up of 45 g of vegetable protein and 5 g of animal protein. Early emphasis was on solving the protein deficiency problem, especially in

[9] Obtained by dividing planned average annual total research by the Plan's projected value added in the agricultural sector during the Plan period.

Table 3. Relative shares of subsectors in GDP (current factor prices) and in total agricultural research, Nigeria.

	1953/54[a]	1954/55[a]	1962–68[a]	1970–74[a,b]	1975–80[c]
Crops (as % of GDP)	34.85	55.21	45.58	33.60	20.78
Crop research (as % of total research)	37.42	59.02	49.04	63.07	57.34
Livestock research (as % of GDP)	6.44	6.29	4.99	NA	NA
Livestock research (as % of total research)	39.40	37.80	10.79	16.71	27.29
Fisheries (as % of GDP)	0.95	0.81	2.59	5.11	NA
Fisheries research (as % of total research)	10.60	11.59	23.47	7.62	11.38
Forestry (as % of GDP)	1.55	1.47	4.44	2.52	NA
Forestry research (as % of total research)	12.58	11.59	16.69	12.61	5.02

[a] GDP figures from F.O.S., Annual Abstract of Statistics (various issues).
[b] GDP figures for agriculture and livestock combined into one.
[c] GDP figure of 20.78% combines agriculture, livestock, fisheries, and forestry.

the south with special stress on veterinary research (animal disease control), expanded fisheries production, and the gradual substitution of cereal production for root and tuber production in the south. This reasoning, based on available nutritional information in the early 1950s, partly accounts for the heavy relative emphasis on veterinary and fisheries research in the early 1950s shown in Table 2. Thus even though fisheries and livestock are not sectorally as important as crop production, relatively high historical research allocations appear to have been based on the need to correct existing nutritional deficiencies.

The Emergence of a National Food and Nutrition Policy

The lack of a national food or nutrition policy for much of Nigeria's agricultural history accounts significantly for the historical lack of a comprehensive national agricultural research policy. The First National Development Plan, 1962–68, had neither an agricultural nor a food policy and it made no distinction between food and export crop research.

The Second National Development Plan, 1970–74, was launched when the nation was still basking in the warmth of having successfully fought a civil war largely with domestic resources. The Plan document was nationalistic in tone, with "self-reliance" and "self-sufficiency" objectives occupying centre stage. The Plan had an eloquent statement of agricultural sector objectives and the food problem and made specific allocations for food crop research. The renewed emphasis on food crop research was further amplified in the Third National Development Plan, 1975–80. These new national concerns were translated into a major institutional reorganization in 1975 when the federal government took over all existing agricultural research institutes and created new ones.

Current Situation

Federal government annual agricultural research allocations are made to the research institutes and are not made on a commodity basis. The relative emphasis placed on crops, livestock, forestry, and fisheries can only be inferred from the statutory research responsibilities assigned to each research institute.

Allocations to Food Crops Research Institutes

Federal government annual allocations to food crops research averaged ₦ 27 381 950 during 1976/77–77/78, representing 40.27% of all federal government annual allocations to agricultural research during the period (Table 4). Table 5 classifies the allocations into commodity groups.

Federal allocations to cereals, grain legumes, seeds, and nuts averaged ₦ 20 526 250 per year during 1976/77–77/78, representing 30.23% of all federal government allocations to agricultural research during the period. The corresponding figures for roots and tubers were ₦4 792 900 and 6.8%, respectively, while those for citrus, fruits, and vegetables were ₦2 062 800 and 3.03% respectively.

Cereals are the most widely grown group of crops and, given current technology levels, employ the largest number of farm workers. They also span the largest ecological area of the country and are therefore important for regional development policy objectives. The grain legumes are the next most important in terms of land area. These are then followed by roots and tubers and seeds and nuts (Table 6).

Percentage research resource allocations to cereals and grain legumes are warranted by their importance in land use, whereas percentage research resource allocations to roots and tubers clearly exceed their relative importance in land areas. However, the position of roots and tubers should be interpreted with caution because they may be more efficient producers of calories per unit of land area. Other commodity groups cannot be evaluated with this land-use criterion because no data exist.

Nutritional Importance of Commodity Groups

To employ this as an evaluative criterion, it becomes necessary to examine research resource allocations to other food items.

Table 4. Federal government budget allocations to food crops research institutes, 1976/77 and 77/78.

Research institute	Commodities for research	1976/77		1977/78	
		Naira	% of allocation to all res. inst.	Naira	% of allocation to all res. inst.
Food Crops					
NCRI	Maize, rice, grain, legumes, sugarcane	6 635 500	12.50	10 789 000	13.02
NIHORT	Citrus, fruits, vegetables	1 581 600	2.98	2 544 000	3.07
NRCRI	Yams, cassava, cocoa, yam, sweet potatoes, Irish potatoes	3 205 000	6.04	6 280 800	7.70
IAR	Sorghum, millet, wheat, barley	6 500 000	12.24	8 528 000	10.29
IAR&T	Cereals and grain, legumes	3 000 000	5.65	5 600 000	6.76
Subtotal		20 922 100	39.40	33 841 800	40.84
Livestock					
NITR	Cattle	2 500 000	4.71	4 560 000	5.50
NAPRI	Cattle, sheep, goats pigs, poultry	1 300 020	2.45	2 640 000	3.19
NVRI	Cattle	5 090 450	9.59	7 472 360	9.02
LRIN	Leather, hides	1 557 970	2.93	3 477 576	4.20
Subtotal		10 448 440	19.68	18 149 936	21.90
Fisheries					
LCRI	Fisheries, irrigated crops	450 400	0.85	2 520 320	3.04
KIRI	Fisheries, irrigated crops	1 553 350	2.93	3 692 000	4.45
NIOMR	Fisheries, irrigated crops	1 510 210	2.84	4 629 768	5.59
Subtotal		3 513 960	6.95	10 842 088	13.08
Tree Crops					
CRIN	Cocoa, coffee, kola, cashew	4 001 000	7.53	5 133 200	6.19
RRIN	Rubber	3 046 000	5.74	1 755 680	2.12
NIFOR	Oil palm, coconut, raphia, dates				
FRIN	Forests	4 611 200	8.68	7 480 976	9.03
Subtotal		16 958 220	31.93	19 615 288	23.55
Grand total		53 103 320	100.00	82 878 592	100.00

Table 5. Federal government allocations to agricultural research by commodity groups relative to their nutritional importance, Nigeria.

	% of allocation to all research institutes[a]		Average contribution to total calorie availability[b] (1972–74) (%)		Average contribution to total protein availability (1972–74) (%)	
	1976/77	1977/78				
Cereals			49.83		42.59	
Seeds and nuts	30.39	30.06	14.82	72.78	23.75	76.12
Pulses (cowpeas)			4.03		9.78	
Sugar			4.10		0	
Vegetables	2.98	3.07	0.81	2.46	3.18	3.97
Fruits			1.65		0.79	
Livestock	19.68	21.90	2.39		5.30	
Fishery	6.95	13.08	0.38		1.94	
Palm kernel/oil	9.98	6.21	2.08		—	
Roots/tubers	6.04	7.70	18.61		11.90	

[a] For the derivation of the first two columns, see Table 4.
[b] Total calorie/protein availability refers to total calorie/protein supplies from domestic production/output.

Table 6. Research allocations to crops compared with land area under crops, Nigeria.

	Share of federal government allocations to all food crop research (%)		Share of commodity group in total hectarage in food crops (%)	
	1976/77	1977/78	1965/66–1969/70	1970/71–1974/75
Cereals, grain, legumes, seeds, and nuts[a]	77.12	73.62	88.48	87.69
Roots and tubers	15.32	18.32	7.9	8.2

[a] Allocations are made to research institutes with statutory commodity research responsibilities. It is not possible to disaggregate the data by commodity class.

Federal government allocations to livestock research during 1976/77–77/78 averaged ₦ 14 299 188 annually, representing an annual average of 20.78% of all federal government allocations to all agricultural research during the period.

Federal government allocations to fisheries research averaged ₦ 7 178 024 during 1976/77–77/78. This represents an annual average of 10.02% of all federal allocations to all agricultural research during the period.

Federal allocations to tree crop research averaged ₦ 17 937 244 per annum during 1976/77–77/78, representing an average of 27.74% per annum during the period.[10]

---

[10] Note that allocations to NIFOR have been grouped with foods because all palm oil is now domestically consumed.

Research Allocations and Nutritional Importance of Crops

From Table 6 it is possible to compare relative financial allocations for commodity research with the nutritional importance of the commodity group. Together, cereals, seeds and nuts, cowpeas, and sugar contributed an average of 72.78% of the country's domestic supply of calories during 1972–74; they also supplied 76.12% of the domestic availability of protein during the same period. The relative share of cereals, seeds and nuts, cowpeas, and sugars in federal government budget allocations shown in Table 5 is warranted by the dominance of these crops as sources of calories and protein.

Whereas seeds and nuts contributed over 23% of total (domestic) protein availability during the period, it is pertinent to note that these crops were not even covered in the Research Institute (Establishment) Order, 1975. It is only with the subsequent funding control that the FMST has come to exercise over IAR that these seeds and nuts are financed by the federal government. There is a need to increase the research emphasis on oil seeds and nuts as well as grain legumes.

Though roots and tubers do not provide as much calories and protein as cereals and grain legumes and are also not efficient sources of calories and protein per unit of labour (the scarce resource), they are nonetheless consumed by large segments of the population — especially the low income group. On the grounds that large consumer surpluses could be conferred on these poor people, a relatively high priority should be accorded root and tuber research.

Research priorities could be assigned on the basis of those crops that economize on some scarce resources. In Nigeria and many other African countries, land is not scarce, though soils are often poor. Ideally, research should emphasize those crops that are most efficient in producing calories or protein per unit of labour, which is currently a scarce resource.

Research Resource Allocation at the Institute for Agricultural Research (IAR)

Only the IAR provided disaggregated research resource allocation data (Table 7). The relative allocation of scientific man-years ranges from an annual average of 4.15% of total scientific man-years in all IAR programs for grain legumes to 7% for groundnuts and oilseeds, with Kano Station serving mainly as a groundnut station. Among the general research programs, the largest share of all scientific man-years goes to socioeconomic research (an annual average of 21.47% of all scientific man-years during 1975/76–77/78). This was followed by soil fertility and soil nutrition programs (13.43%). Agricultural mechanization received 8.91% and irrigation research had 11.25%.

The allocations of scientific man-years among commodities do not accord with the relative national importance of commodity groups as sources of calories and protein. Roughly the same patterns of allocation are obtained for the financial resources though the relative allocation to cereals is slightly higher.

Input Research

Most agricultural research in Nigeria has concentrated largely on commodities: crops, livestock, fisheries, and forest products. The little input research that has occurred has only been of the nature of ''intermediate'' or ''lower level'' research that is necessary for some other commodity-oriented ''final'' or higher level research. Thus, soil fertility studies have been carried out on an ad hoc basis largely in response to the needs of crop agronomists working on specific crops. There is no coordinated research program on soils within a national framework, i.e., research that views land as a national resource requiring comprehensive basic knowledge of nutrients and having potential as a basis for rational use and management in crop and livestock production. In fact, none of the existing agricultural research institutes has a specific mandate to study soils from a national perspective.

The need for rational soil management research from a national perspective arises also from the fact that the private costs and social costs of rotational bush fallowing differ. To the extent that private farmers' costs of rotational bush fallowing are consistently lower than the costs to the whole society, there is a tendency to overuse the land and deplete soil nutrients faster than would have been the case if private costs of soil exploitation were equal to the social costs of soil exploitation. This lack of major institutional responsibility for soil research at the national level creates gaps in research knowledge especially with respect to the optimal spatial allocation of agricultural production patterns within the country.

Farm Labour Research

Labour is still the most important input in Nigerian agriculture. Yet, little research has been carried out on various aspects of farm labour in traditional agriculture: labour utilization and labour profiles in different crops and ecological zones; the structure of farm labour markets; the link between rural farm labour markets and nonfarm labour markets; supply and demand relations in farm labour, etc. There have been isolated, but very limited, insights into farm labour utilization in the few farm management studies that have been carried out. Yet no national study has ever been commissioned to examine various aspects of farm labour utilization problems. However, macroeconomic and agricultural sector projects are being embarked upon that continue to run into farm labour shortages and difficulties, especially during the peak-season farming operations. The present organizational setup of agricultural research institutes assigns no responsibility to any institute for research into socioeconomic research. The absence of socioeconomic research is a serious gap because agricultural production is in the hands of millions of farmers operating for profit. Socioeconomic research should receive high priority because adoption of agricultural research results depends on their profitability.

There is urgent need for research into the most appropriate forms of machines that will substitute for labour as a source of farm power. The allocation

Table 7. Allocations of research resources to various research programs, IAR.[a]

	1975/76		1976/77		1977/78		1978/79		Mean percentage (1975/76–1977/78)	Share of financial resources for all programs 1978/79
	Senior scientific man-years	% of all scientific man-years	Senior scientific man-years	% of all scientific man-years	Senior scientific man-years	% of all scientific man-years	Senior scientific man-years	% of all scientific man-years		
Cereal program					14.0	6.56	NA	NA	6.56	5.17
Cotton and fibres	8.025	11.80	7.025	8.71	7.15	3.35	8.15	5.26	7.95	
Groundnuts and oil seeds	6.1	8.97	8.85	10.97	8.05	3.77	9.75	6.29	7.90	2.27
Grain legumes program	4.4	6.47	3.4	4.21	3.75	1.76	3.75	2.42	4.15	1.37
Agric. Res. Station, Kano	0.5	0.74	0.5	0.62	8.0	3.75	8.0	5.16	1.70	—
IRS, Ngala	0.5	0.74	0.5	0.62	4.0	1.87	4.0	2.58	1.08	—
Horticultural groups	5.1	7.50	5.1	6.32	6.7	3.14	8.35	5.39	5.65	2.18
Cropping systems	5.0	7.35	5.0	6.20	8.1	3.80	14.6	9.42	5.78	3.78
Socioeconomic and extension	18.5	27.20	21.5	26.65	22.5	10.55	21.5	13.87	21.47	4.56
Food science and tech. tech. grasshopper project	0.9	0.13	2.8	3.47	3.05	1.43	6.05	3.90	1.68	—
Agric. mechanization	5.5	8.09	10.5	13.02	12.0	5.62	5.0	4.34	8.91	3.35
Irrigation research					24.0	11.25	24.0	15.48	11.25	3.23
Termites projects	3.0	4.41	3.0	3.72	4.05	1.90	4.05	2.61	3.34	2.35
Irrigation research station, Bakura					4.0	1.87	3.0	1.94	1.87	—

[a] Source: Idachaba, F.S. 1980. Agricultural research policy in Nigeria. International Food Policy Research Institute, Washington, D.C.

of research resources to farm mechanization at IAR has been respectable: during 1975/76–1977/78, an annual average of 8.91% of all senior scientific man-years in IAR research was allocated to farm mechanization. It was proposed to spend an average of 3.35% of the 1978/79 budget on farm mechanization.

The relative allocation of research resources to soil fertility and nutrition research at IAR has been much more impressive than the national picture. During 1975/76–1977/78, an annual average of 13.43% of all scientific man-years in IAR was allocated to this program. It was also proposed to spend 5.95% of the IAR budget on this program in 1978/79. The importance of this program, like the socioeconomic program, for all crops, cannot be overemphasized.

Irrigation Versus Rain-Fed Agricultural Research

Most of the agricultural research in Nigeria has concentrated on rain-fed agriculture. The neglect of irrigation agricultural research can be traced to the historical neglect of a national policy on irrigation agriculture. Changes in national economic objectives have, however, been translated into agricultural research priorities in recent years. Such changes, evident in the 1970–74 and 1975–80 plans, include the objectives of: a stable national economy; minimum nutritional standards for all citizens; self-reliance, especially in its operational form of self-sufficiency in food; etc. The objective of a stable national economy calls for policies that will insure the national economy against the kinds of discontinuity created by droughts and severe environmental stress, as was the case in the 1973/74 season. Irrigation agriculture has since then received new emphasis to insure the national economy against erratic and uncontrollable fluctuations. The minimum calorie objective calls for irrigated production of cereals, dry season vegetables, etc., that will minimize nutritional deficiencies during the lean season. Self-reliance and self-sufficiency objectives transform into agricultural policies to produce irrigated wheat, vegetables, and other crops to cut down on the nation's dependence on others for these commodities. These new policy objectives have resulted in a major emphasis on irrigation agriculture, as reflected in the almost simultaneous creation of 11 River Basin Development Authorities. The new emphasis on irrigation agriculture is shown in the 1975–80 Plan. Federal government allocations in millions of naira (with percentages for all agriculture in parentheses) to the different subsectors were: agriculture 765.028 (45.68%); irrigation 535.086 (31.95%); livestock 284.019 (16.96%); fisheries 54.560 (3.26%); and forestry 36.130 (2.16%). The irrigation figure is an underestimate because several projects under irrigation were merged with projects under agriculture. At the institute level, irrigation research received an annual average of 11.25% of all scientific man-years in 1978/79. It was also proposed to spend 3.23% of the research budget for 1978/79 on irrigation research. Given the importance of irrigation in regional development as well as in stabilizing agricultural production, it is suggested that the amount allocated to irrigation be raised substantially. It is not sufficient to simultaneously launch 11 River Basin Development Authorities. For these to achieve maximum impact and be translated into crop production, irrigation research must be accorded high priority. Otherwise the enormous resources employed in these authorities could go to waste.

Criteria for Determining Agricultural Research Priorities: A Synthesis

Two preliminary observations are necessary. One, allocative criteria are essentially country-specific, reflecting the peculiar opportunities and constraints of a country, and are not necessarily applicable to other countries. Two, allocative criteria even for a given country are essentially dynamic in nature and respond to changes in objectives, opportunities, and constraints.

(1) Foreign Exchange Contribution of Agricultural Commodity

A working rule could be: allocate research resources to commodities in direct proportion to their contributions, either from savings in dollars through import substitution or dollar earnings through exports. But this is only the first step and appears workable when there is a crop research budget to be allocated to commodities with foreign exchange contributions. It leaves unanswered the issue of how much to allocate to foreign exchange earners/savers and how much to domestically produced and traded agricultural commodities that do not enter international trade. Most Third World countries intially concentrated research on foreign exchange earners, almost to the neglect of food crops. The salient features of national research systems at this initial stage were: disproportionately larger financial allocations to export crop research; export crop research bias in the allocation of research time and research experiments; export crop bias in the allocation of research personnel; and locations of research institutes and substations largely in the export crop growing areas (e.g., the location of IAR and its substations largely in the cotton and groundnut belt; the location of Cocoa Research Institute of Nigeria in the cocoa belt; NIFOR in the palm belt; and the classic historic error of locating a Research Station at Moor Plantation in the mistaken belief that its environs would be suitable for cotton production).

The relevant weights for allocation to export crop research in this working rule are the share of agricultural exports in total export earnings and the share of agriculture in GDP.

(2) Fiscal Role of Crop

In an ideal world, there should be no policy-induced distortions in the form of commodity taxes/levies by statutory export monopsonies. The real world is filled with marketing boards and the like and the taxes they levy on export crop producers are an important source of government revenue. To the extent that these monopsonies have become institutional realities in these countries, we have a second-best problem in which efforts must then be made not to kill the goose that lays the golden egg. Within this constrained format, the working rule is: allocate research resources to crops in direct proportion to their relative importance in government revenues. The objective of this working rule is to expand, through appropriate research support, the production and the public revenue base of the crop. In many cases, the statutory monopsonists themselves provided grants to the crop research institutes.[11]

(3) Value of Production

The working rule here is to allocate research resources in direct proportion to the relative actual and potential value of production crops. In properly functioning markets, the areas under the aggregate demand curves for crops should reflect their social valuation or benefits, which should then serve as guides for the derived demands for agricultural research in the respective crops.

A particular variant of this criterion is the availability of assured markets for the intermediate and final outputs of research. The availability of European markets provided sufficient motivation for public research managers and researchers alike to boldly allocate substantial resources to export crop research, especially during the colonial era. Traditional food markets, though relatively efficient in the local context, remained largely segmented. The subsidiary working rule here is: research resource allocations should be directly proportional to the availability of assured markets.

(4) Value of Urban Consumption

Urban/city dwellers in developing countries, dependent as they are on others for their food supplies, are quick to revolt whenever their food lifelines are threatened. Thus, foods with higher per capita urban consumption (probably reflecting a per capita income demand elasticity differential) will tend to command higher agricultural research priority. This is because of the social costs of urban riots and political unrest. In Nigeria, for example, colossal amounts are being provided to boost rice production and to attack major research bottlenecks; in fact N103.6 million has just been allocated for boosting rice production alone. Yet rice does not compare with sorghum or millet in terms of value of production.[12] Also, huge sums are being spent to develop large-scale irrigation schemes, one of the aims of which is to promote the production of dry season wheat in Nigeria. Urban per capita consumption of wheat in Nigeria is far higher than rural per capita consumption. Breeding research effort at IAR, Samaru, is concentrating on wheat selections for yield, quality, and resistance to disease and lodging under Nigerian conditions.

---

[11] In Nigeria, 7.5% of operating revenue of produce marketing boards was statutorily to be expended on research. Grants for agricultural research by the Regional Marketing Boards during 1955–61 were: Western Nigeria N 10 million (mainly cocoa research and extension); Northern Region N 5.6 million (mainly to Samaru Research Station with N 3.2 million for general research while N 2.4 million was specifically for cotton development). The World Bank had earlier recommended that the normal recurring costs of the West African Institute for Oil Palm Research (WAIFOR) would be met from an endowment to which the Nigerian Oil Palm Produce Marketing Board contributed 82%. The Cocoa Marketing Board also provided funds for WAIFOR. In 1953/54, the Cocoa Marketing Board provided N 474 000 for cocoa and soil survey. In Northern Nigeria, the Marketing Board had provided N 226 000 for cotton development by 1953/54.

---

[12] Lowland rice breeding research priorities at Moor Plantation and Badeggi are on the development of new high-yielding varieties that: are resistant to rice blast (*Pyricularia oryzae*) and other diseases; are responsive to fertilizers; are quick maturing; and economize on scarce water resources. Upland rice breeding priorities, in addition, concentrate on varieties that are resistant to leaf spot disease (*Heminthesporium oryzae*) with narrow leaves and shoot, stiff straw and long grain. A new major research priority is rice processing because of the demonstrated income elastic demand for well processed imported long grain rice from the United States.

(5) Regional Development

No large sections of the country should be left behind in the development process. The implied working rule from this proposition is: research resources should be allocated in direct proportion to the relative shares of land areas under crops. The larger the ecological spread of a crop within the country, the higher should be its research emphasis for purposes of balanced regional development.

(6) Employment Generation Potential

Many food-deficit countries are plagued with large pockets of rural unemployment and underemployment even after one has adjusted for the seasonal factor in farm-labour demand. With only modest growth in urban industrial employment, agricultural research priorities should centre around crops and techniques with gainful employment opportunities that will absorb the large pockets of the rural unemployed and underemployed. In such countries, research resources should be allocated in direct proportion to the direct and indirect labour intensities of crops and new techniques of production and distribution. Social returns to such research are high because it makes possible the productive social utilization of resources that would have otherwise been idle.

Put in another way, research priority should concentrate on crops with large labour-intensive value added. In Nigeria, groundnut and textile mills have had to close periodically and lay off thousands of workers either because of disastrous outbreaks of rosette disease in groundnuts (as was the case in 1975) or because of poor rainfall distribution in the cotton growing areas. In these instances, research that successfully breeds rosette-resistant groundnuts or cotton that is resistant to moisture stress will indirectly promote off-farm employment in the factories at Kano and Kaduna.

(7) Politically Visible Crops

Social reactions, especially in the urban areas, to shortfalls in supplies vary from food item to food item. Among the politically most visible are convenience foods (bread), sugar, milk, and rice. The world market for sugar is particularly volatile and a country may not wish domestic availability of sugar to fluctuate with world market conditions. The more strategic a crop is in terms of political visibility, the higher the research priority that will be assigned to the crop. This is not merely a guideline of self-preservation for the political elite. It reveals the responses of research management to articulate urban pressure groups.

(8) Nutritional Significance of Crop

The greater the significance of a crop or commodity group as a source of calories or protein, the higher the research priority that should be accorded to the crop or the commodity group. The nutrition weights to be applied for research resource allocation could be based on a particular base period or base year.

(9) Value-Added and Import-Substitution Industrialization

Research resources should be allocated in direct proportion to the contributions of crops as raw materials for import-substitution industrialization. This is closely related to the foreign exchange contribution criterion. In addition, this criterion states that research resource allocations should be proportional to the degree of value added.

(10) Narrowing of Wealth and Income Inequality

In order that large segments of the rural poor will not be left behind in the development process, high priority should be accorded to research that will liberalize accessibility of the rural majority to wealth sources and income streams. The several strands of this criterion include: (1) research priority on new inputs that are within economic reach of the poor rural majority; (2) research priority on the wealth and income distribution consequences of new technologies put out by the research system; (3) research priority on the sources of existing rural wealth and income inequalities and how new technologies may exacerbate such inequalities; and (4) research priorities on positive measures for narrowing rural wealth and income inequalities.

(11) The Foods that the Poor People Eat

In attempts to uplift the welfare of the poor, research priority could be accorded those foods with income elastic demands at low income levels. This will ensure that growing demands for such foods with income growth can be met at prices that poor people can afford.

(12) Resource Allocations in Input Research

Resource allocations to input research should be in direct proportion to the "importance" of farm inputs in agricultural production. "Importance" could be measured in one of two ways: either the "constraint ranking" of the input or the contribution of the input to output as measured by its output elasticity. Either way, labour ranks first in Nigeria, but it remains the least researched farm input in Nigerian agriculture.

Conflicts in Allocative Criteria

Unfortunately, there are bound to be conflicts between our specified allocative criteria — at least if we try to apply the whole list to a single country. For example, export crops and nutritionally important crops may be competing for the same resources. Many other potential conflicts could be cited.

There are probably two ways out. One is to assign weights to each criterion that are valid for a country's circumstances at any particular point in time. Conflict resolution is then based on relative weights. The other way out is to eliminate the conflict at source, i.e., conflicts between the "higher level" objectives/priorities from which agricultural research priorities have been derived through the sequential transformation process. Still, conflicts may have their origins in the sociopolitical dynamics of the country that reflect relative demands of competing interest groups. Such conflicts may only be resolved through hard political bargaining.

Lags in Agricultural Research Priorities

Four types of lags need brief mention. First is the lag between the existence of a research priority and its recognition as such by both the political and the research leadership. Determinants of this lag include whether or not an explicit statement of higher-level objectives/priorities exists and the time it takes for "higher level" priorities/objectives to be transformed into "lower level" research priorities. Second is the lag between the recognition of a research priority and the formulation of a research program. Third is the lag between the formulation of a research program and the emergence of a research solution. Determinants of this lag centre on those factors that determine agricultural research productivity: stable and consistent funding; the quality and stability of research staff; supporting technical personnel; supplies and materials; etc. The final lag is that between a research priority solution and mass impact, on farmers and consumers alike. A major policy concern in national research management is how to shorten these lags.

A Methodology for Establishing Priorities for Research on Agricultural Products[1]

Luis J. Paz[2]

A method is being developed in Peru by the Fundacion para el Desarrollo Nacional[3] to assign economic resources to research in agricultural and livestock products to maximize the development of the country, particularly the rural sector. This paper presents a summary of the preliminary results of the study, which seeks to establish research priorities based on specific products. The method tries to maximize objectivity in establishing the criteria upon which to determine commodity research priorities. The next step will be to establish product priorities by region and type of research for each product in each region. In each of the regions, farmers' associations will be able to influence these research priorities by providing additional funds for research on specific products. The final results of this process can also be used to establish research priorities for resources (water, soil, etc.) and activities (disease control, seed production, etc.).

Research priorities were determined for 53 agricultural products and 16 livestock products on which research might be carried out in Peru. Priorities were established for each product on the basis of its contribution to government objectives. Thirteen objectives were used, with each objective given a weighting in terms of its relative importance to the country. Each product was then assessed in terms of its contribution to one objective at a time. The weights given to one product for each of the 13 objectives were then added to give an overall priority rating for that product.

The 13 national objectives used in this study are listed in Table 1, along with the relative importance attached to each. Relative weights were assigned by a panel of six judges who are experts in the field of agriculture in Peru. After an initial rating and discussion among the judges of the importance of these objectives, each of them assigned a total of 100 points to the 13 objectives. An average was then calculated to give the final weightings shown in Table 1.

The total weight assigned to each objective (e.g., 11.97 for gross value of total production) was then

Table 1. Government criteria used to assess agricultural product priorities.

Criteria	Panel-assigned weights
Quantitative	
Gross value of total production	11.97
Aggregated value per hectare	10.97
Employment generation (labour per hectare)	10.57
Cultivated land or fixed capital	8.18
Proportion of the family basket	7.58
Saving of foreign currency through import reductions	7.02
Importance as imports of principal industries (gross value of imports/gross value of industries)	5.98
Value of national imports used/gross value of the agricultural product	5.09
Nutritional value	2.55
Qualitative	
Importance of the incorporation of new land	9.88
Potential for productivity increment	9.28
Possibilities for industrialization	6.54
Potential production due to ecological diversity	4.39
Total	100.00

---

[1] This is a preliminary report presented for discussion purposes. The methodology is undergoing revision and is expected to be finalized by the end of 1981.

[2] Luis F. Villaran 383, Lima 27, Peru.

[3] The Fundacion para el Desarrollo Nacional is a non-profit institution created in 1970 with the objective of getting maximum utilization from the installed capacity of the National Agricultural University of La Molina.

Table 2. Agricultural product research priority coefficients.

Product	Gross value of total production	Aggregated value per hectare	Employment generation (labour per hectare)	Importance of the incorporation of new land	Potential for productivity increment	Cultivated land or fixed capital	Proportion of the family basket	Saving of foreign currency through import reductions	Possibilities for industrialization	Importance as imports of principal industries (gross value of imports/gross value of industries)	Value of national imports used/gross value of the agricultural product	Potential production due to ecological diversity	Nutritional value	Total
Cotton	0.444	0.210	0.126	0.350	0.305	0.267	—	0.335	0.228	0.599	0.081	0.157	0.099	2.972
Sugarcane	0.417	0.287	0.138	0.305	0.305	0.243	0.274	—	0.247	0.535	0.129	0.137	0.098	2.830
Coffee	0.391	0.417	0.157	0.326	0.326	0.261	0.223	—	0.154	0.441	0.006	0.047	0.022	2.788
Cocoa	0.130	0.375	0.157	0.276	0.300	0.137	0.073	—	0.195	0.346	0.024	0.041	0.094	2.148
Tea	0.121	0.408	0.176	0.258	0.300	0.095	0.162	0.320	0.083	—	0.074	0.044	0.019	2.060
Tobacco	0.182	0.188	0.061	0.266	0.145	0.083	—	0.372	0.116	0.567	0.146	0.098	—	2.224
Fruits														
Citrus	0.365	0.123	0.115	0.290	0.182	0.226	0.262	—	0.195	0.047	0.166	0.133	0.036	2.140
Pomaceous (apples and similar)	0.321	0.309	0.069	0.207	0.269	0.160	0.201	0.290	0.195	—	0.048	0.117	0.044	2.230
Fruits with stone	0.260	0.342	0.076	0.155	0.145	0.131	0.123	—	0.195	—	0.033	0.114	0.052	1.676
Gourd family	0.226	0.112	0.322	0.143	0.145	0.059	0.056	—	0.033	—	0.066	0.083	0.025	1.270
Plantain	0.382	0.320	0.084	0.363	0.182	0.237	0.235	0.246	0.116	—	0.017	0.121	0.057	2.360
Pineapple	0.208	0.079	0.222	0.266	0.091	0.107	0.061	0.224	0.195	—	0.011	0.130	0.034	1.628
Papaya	0.200	0.123	0.199	0.302	0.121	0.125	0.078	—	0.195	—	0.006	0.127	0.031	1.507
Avocado	0.269	0.331	0.023	0.225	0.091	0.166	0.190	—	0.033	—	0.017	0.124	0.050	1.519
Olive	0.278	0.243	0.049	0.172	0.182	0.077	0.129	—	0.116	—	0.063	0.010	0.061	1.380
Vine	0.330	0.331	0.176	0.232	0.269	0.178	0.168	—	0.195	0.504	0.037	0.023	0.048	2.491
Prickly pears	0.096	0.309	0.016	0.187	0.300	0.071	0.040	—	0.033	—	0.024	0.055	0.046	1.180
Beans														
Kidney bean	0.339	0.006	0.264	0.223	0.229	0.249	0.240	0.253	0.083	—	0.163	0.153	0.067	2.269
Chick-peas	0.061	0.007	0.264	0.132	0.229	0.101	0.067	—	0.083	—	0.170	0.038	0.082	1.234
Horsebean	0.148	0.199	0.264	0.161	0.229	0.148	0.145	—	0.083	—	0.135	0.035	0.071	1.618

Lentils	0.035	0.024	0.264	0.129	0.229	0.042	0.140	—	0.083	—	0.135	0.032	0.084	1.197
Garden peas	0.191	0.079	0.264	0.121	0.229	0.214	0.111	0.283	0.116	0.236	0.126	0.112	0.080	2.762
Lima beans	0.234	0.035	0.264	0.155	0.229	0.220	0.095	—	0.116	—	0.140	0.095	0.065	1.648
Vegetables														
Leafy plant	0.252	0.221	0.360	0.046	0.057	0.142	0.218	0.298	0.033	—	0.174	0.079	0.040	1.920
Root	0.165	0.232	0.333	0.046	0.057	0.065	0.212	—	0.033	—	0.181	0.076	0.038	1.438
Fruit-bearing plants	0.174	0.145	0.291	0.051	0.091	0.053	0.156	0.264	0.033	0.047	0.071	0.073	0.033	1.482
Legumes	0.078	0.287	0.191	0.051	0.057	0.030	0.101	—	0.033	—	0.100	0.070	0.029	1.027
Bulb	0.347	0.221	0.352	0.137	0.182	0.172	0.257	0.238	0.116	0.157	0.146	0.109	0.055	2.489
Other vegetable	0.373	0.101	0.237	0.057	0.091	0.231	0.246	—	0.195	—	0.116	0.160	0.073	1.920
Gourd family	0.217	0.232	0.391	0.071	0.091	0.113	0.207	—	0.033	—	0.085	0.067	0.028	1.535
Tomato	0.286	0.353	0.398	0.078	0.182	0.154	0.251	—	0.195	—	0.044	0.092	0.024	—
Cereals														
Corn	0.434	0.146	0.210	0.351	0.323	0.308	0.229	0.365	0.154	0.409	0.116	0.162	0.090	3.297
Rice	0.408	0.167	0.345	0.372	0.269	0.255	0.285	—	0.168	—	0.122	0.149	0.086	2.626
Wheat	0.356	0.027	0.210	0.137	0.229	0.279	0.290	0.387	0.154	0.315	0.192	0.055	0.088	2.713
Barley	0.139	0.020	0.017	0.229	0.323	0.291	0.268	0.357	0.169	0.472	0.185	0.064	0.078	2.612
Quinoa	0.087	0.090	0.126	0.187	0.343	0.196	0.089	—	0.154	0.283	0.035	0.020	0.075	1.705
Minor cereals	0.156	0.020	0.038	0.096	0.323	0.181	0.173	0.350	0.154	—	0.188	0.089	0.077	1.848
Root Crops														
Potato	0.453	0.167	0.306	0.223	0.343	0.303	0.279	0.275	0.154	—	0.159	0.113	0.069	2.874
Cassava	0.373	0.134	0.049	0.333	0.300	0.285	0.151	—	0.154	—	0.030	0.102	0.059	1.970
Sweet potato	0.295	0.101	0.306	0.254	0.145	0.190	0.134	—	0.116	0.236	0.153	0.106	0.054	2.090
Ollocus (yellow potato)	0.243	0.068	0.306	0.157	0.091	0.202	0.106	—	0.033	—	0.153	0.016	0.042	1.417
Oilseeds														
Oil palm	0.017	0.287	0.107	0.363	0.182	0.024	0.179	0.231	0.237	—	0.091	0.013	0.092	1.823
Soybean	0.043	0.046	0.230	0.261	0.023	0.036	0.184	0.380	0.237	—	0.177	0.086	0.063	1.766
Oleaginous	0.113	0.178	0.099	0.086	0.020	0.089	0.196	0.342	0.237	0.157	0.091	0.061	0.092	1.761
Spices														
Pepper	0.052	0.265	0.379	0.086	0.020	0.006	0.084	0.313	0.034	0.047	0.107	0.007	—	1.400
Oregano	0.026	0.263	0.092	0.003	0.057	0.012	0.050	—	0.034	—	0.107	0.029	—	0.673
Various spices	0.009	0.263	0.333	0.007	0.020	0.048	0.117	0.328	0.034	0.157	0.107	0.051	—	1.474
Fibres														
Jute	0.104	0.362	0.368	0.302	0.121	0.119	—	0.305	0.221	0.378	0.041	0.004	—	2.325
Cocaine	0.304	0.373	0.145	0.014	0.020	0.208	0.045	—	0.033	—	0.024	0.026	—	1.192
Flowers	0.069	0.176	0.379	0.042	0.260	0.018	—	0.264	0.034	—	0.074	0.140	—	1.465
Pyrethrum	0.003	0.384	0.176	0.021	0.020	—	—	—	0.154	—	0.074	0.001	—	0.833
Pasture/Forage														
Alfalfa	0.426	0.395	0.030	0.276	0.182	0.273	—	—	0.083	0.047	0.055	0.146	—	1.913
Cultivated pastures	0.400	0.395	0.016	0.344	0.269	0.297	—	—	0.033	—	0.055	0.165	—	1.974
Total	11.970	10.970	10.570	9.880	9.280	8.180	7.580	7.020	6.540	5.980	5.090	4.390	2.550	100.000

Table 3. Product weights for value of production criterion.

	Ranking	%	Value
Potato	52	3.774	0.452
Cotton	51	3.701	0.443
Corn	50	3.628	0.434
Alfalfa	49	3.556	0.426
Sugarcane	48	3.483	0.417
Rice	47	3.411	0.408
Cultivated pastures	46	3.338	0.400
Coffee	45	3.266	0.391
Plantain	44	3.193	0.382
Cassava	43	3.120	0.373
Citrus fruits	42	3.048	0.365
Wheat	41	2.975	0.356
Fruit-bearing plant bulbs	40	2.903	0.347
Kidney bean	39	2.830	0.339
Vine	38	2.758	0.330
Pomaceous	37	2.685	0.321
Vegetable	36	2.612	0.313
Cocaine	35	2.540	0.304
Sweet potato	34	2.467	0.295
Tomato	33	2.395	0.287
Olive	32	2.322	0.278
Avocado	31	2.250	0.269
Fruits with stone	30	2.177	0.261
Leafy plants	29	2.104	0.252
Ollocus (yellow potato)	28	2.032	0.243
Lima beans	27	1.959	0.234
Gourds (fruits)	26	1.887	0.226
Gourds (vegetables)	25	1.814	0.217
Pineapple	24	1.742	0.209
Papaya	23	1.669	0.200
Garden peas	22	1.597	0.191
Tobacco	21	1.524	0.182
Fruit-bearing plants	20	1.451	0.174
Root	19	1.379	0.165
Minor cereals	18	1.306	0.156
Horsebean	17	1.234	0.148
Barley	16	1.161	0.139
Cocoa	15	1.089	0.130
Tea	14	1.016	0.122
Oleaginous	13	0.943	0.113
Jute	12	0.871	0.104
Prickly pears	11	0.798	0.096
Quinoa	10	0.726	0.087
Flower (legume)	9	0.653	0.078
Flowers	8	0.581	0.070
Chick-peas	7	0.508	0.061
Pepper	6	0.435	0.052
Soybean	5	0.363	0.043
Lentils	4	0.290	0.035
Oregano	3	0.218	0.026
Oil palm	2	0.145	0.017
Various spices	1	0.073	0.009
Total	1378		

distributed among the 53 agricultural products in the following manner. Each of the 53 products was given a rank from 52 to 0 according to its relative contribution to value of production (Table 3). The relative contribution of each commodity was then determined by dividing its rank by the sum of all ranks (52 + 51 + ... + 1 = 1378) assigned. These percentages were then multiplied by the weight (e.g., 11.97) given to that objective. A similar procedure was followed for each objective, although the panel of judges had to decide the ranking of agricultural products by subjective assessment for the four qualitative criteria (Table 2).

To establish the priority that should be accorded to each of the products based on the quantitative criteria, statistical data were collected. These included: average gross value of production for 1972–76; aggregate value (salaries, profits, interest, and depreciation divided by gross value of production) per hectare; employment generated (labour required per hectare); cultivated area (average area cultivated between 1972 and 1976); proportion of family basket (percentage of each product in the family basket); saving of foreign currency (annual average saving 1972–76; in some cases the agricultural product was considered, in others, the intermediate industrial product); importance as input for principal industries (gross value of inputs/gross value of industries); relationship between value of national inputs used and gross value of agricultural product (proportion of national inputs used to produce 1 ha of crop); and nutritional value (number of grams of protein, oil, carbohydrates, and vitamins per 100 grams of product). The overall weights given to the 53 agricultural products are shown in Table 2.

Conclusions

Table 2 can be used in different ways to assign economic resources to research priorities. The number of criteria and weight assigned to them can be changed according to the specific objectives of the government and as the statistical data and knowledge about these agricultural and livestock products are improved.

The results have been shown to agricultural experts in both the public and private agricultural sectors, and there is, essentially, agreement on the order given to the products as priorities for research at the national level. However, more work is yet required and the weight given to each criterion is subject to further analysis and discussion.

Much of this work has been done with the collaboration of Amador Merino Reyna, Gustavo Gonzales Prieto, and Efrain Palti Solano to whom the author expresses his acknowledgement.

The System of Resource Allocation to Agricultural Research in Kenya

S. N. Muturi[1]

Kenya has had an organized tradition of agricultural research since the turn of the century. Research services have advanced as the situation has demanded. In Kenya, research serves the development of the country rather than its own right. Most research activities have therefore been in the applied sciences.

The first agricultural research institution, the Scott (now National) Agricultural Laboratories was established in 1903. This was followed by the Coffee Research Services (1908), the Veterinary Research Laboratories (1908), and the National Plant Breeding Station at Njoro (1927). The extensive development of agricultural research services, however, occurred in the forties and early fifties when 16 more major research stations were established under the control of the Department of Agriculture.

The same period saw the creation of five main regional research establishments under the aegis of the East African Community. The research establishments were concerned with agriculture, veterinary science, fisheries, and forestry. The main objective of these research establishments was to concentrate on regional problems and strategic science common to the three East African countries (Kenya, Uganda, Tanzania). The national research establishments were to concentrate on local problems and tactical science.

Other research establishments related to agriculture and land use were established in the early sixties. These included research in wildlife management and conservation, and meteorology. The social sciences have also appeared on the scene, the major development being the establishment of the Institute of Development Studies in 1965.

Various agricultural education institutions were also established to provide scientific human resources for the management of agriculture, which was becoming increasingly complex. Thus, the Egerton Agricultural College was upgraded to a full technician training level in 1952. Local training of laboratory technicians came later with the establishment of the Kenya Polytechnic (1961) and Mombasa Polytechnic (1974). The present University of Nairobi (1970) grew from the original Technical and Commercial Institute (1947), which became the Royal College of Nairobi (1961), having been for a time the Royal Technical College of East Africa (1951). The University of Nairobi has fully fledged Faculties of Science, Agriculture, and Veterinary Medicine through which the majority of the existing research workers have been trained.

Management of Agriculture Research

Government thinking on the need for formal machinery for making and implementing policy for science, technology, and research in all sectors of the economy started immediately after Kenya became independent in 1963. It was noted that there was no centralized responsibility for the formulation of scientific policy. Responsibilities for research policy and management lay with individual departments of Government ministries and there was little coordination between the various agencies. There was, for example, no linkage between veterinary research in the Department of Veterinary Services and animal production research in the Department of Agriculture although the two departments were in the Ministry of Agriculture. The same situation prevailed in research establishments in other sectors.

In 1970, stock was taken of the financial and manpower resources allocated to science and technology (Table 1). The scientific and technical activities were estimated to involve about K£24 million of which K£5 million was allocated to research and development (R&D) and K£19 million to other

---

[1] Science Secretary, National Council for Science and Technology, P.O. Box 30623, Nairobi, Kenya.

Table 1. Financial resources (millions of K£) and manpower resources allocated to science and technology in Kenya, 1970 (from Development Plan 1974–78, Government Printer, Nairobi).

| Science group | Expenditure | | Manpower | | | |
| | | | R&D | | STS | |
	R&D[a]	STS	Sci.	Tech.	Sci.	Tech.
Agriculture	3.57	10.33	371	525	228	4300
Medical	0.56	6.03	55	382	988	2219
Natural	0.53	2.01	68	65	99	316
Industrial	0.40	0.92	30	62	1070	1000
Social	0.08	0.42	45	—	22	—
Total	5.14	19.71	569	1034	2407	7835

[a]R&D = research and development; STS = scientific and technical services; Sci. = scientists; and Tech. = technicians.

scientific and technical services (STS). Of the R&D budget, agricultural research accounted for 70%, natural and medical sciences 10% each, and industrial sciences 7%. Some 3000 scientists and 8800 technicians were involved.

The estimated total gross expenditure on R&D was equivalent to 0.91% GDP. The Government funded 75% of R&D and 81% of STS, representing 2.30% and 11.17% of the national budget, respectively. The remainder was funded by higher education and the private sector. Fundamental research was largely undertaken in the University of Nairobi, representing approximately 1.7% of the total research expenditure.

With this magnitude of resources devoted to R&D and STS, the Government felt the need to establish a system for policy formulation and execution of scientific and technological activities. It was not, however, until 1977 that through an Act of Parliament,[2] the Government established the machinery for advising itself on all matters relating to scientific and technological activities, the research necessary for the proper development of the country, and for the coordination of research and experimental development. This machinery comprises: (1) the National Council for Science and Technology (NCST); (2) the Sectoral Scientific Advisory Research Committees (ARCs); and (3) the Statutory Research Institutes (SRIs).

This machinery establishes a series of circuits that connect the requirements of socioeconomic de-

velopment with the utilization of science and technology for development. The machinery links the executive level with the management and policymaking level and the execution and operations level. They provide lines through which ideas and policy flow. The three functional levels that have been established are: (1) policy and strategy; (2) management and tactics; and (3) execution and operations.

The first functional level (policy and strategy) comes under the review of the National Council for Science and Technology, which is legally required, among other things: (1) to advise the Government on a national science policy, including general planning and the assessment of the requisite financial resources; (2) to advise the Government on the overall financial requirement for the implementation of the national science policy and on the disbursements to the agencies concerned; and (3) to review generally and advise on the program and budgets for the promotion of research and related scientific activities proposed by ministries and ensure that they are in harmony with the national science policy.

This science policy circuit (NCST) brings eminent scientists (appointed members) and the policymaking officers of the Government (Permanent Secretaries) together in a socioeconomic forum to debate the use of science and technology for development. The scientists provide the technical inputs and ideas of promising developments. The policymakers know the political and socioeconomic constraints at the highest level. The NCST is so constituted that a concensus among its members as to a course of action has real authority and places an onus of implementation on those concerned, especially its members (Permanent Secretaries). Where, however, NCST considers that the matter requires Cabinet approval, it places before it the necessary recommendations for approval. This applies to such matters as the national budget for science and technology, legislative requirements, and subjects with significant political constraints. The NCST is largely self-advisory on this matter because its ex-officio members know when Cabinet or ministerial endorsement is required or desirable.

In advising on the allocation of resources, the NCST has responsibility for ensuring that all science groups are catered for in both R&D and STS. Government allocations to various science groups in the 1979–80 fiscal year are indicated in Table 2.

The Government's expenditure of 70% of the budget on agricultural sciences is in keeping with the role the sector contributes to the GDP (about 35%) and the infrastructure that agricultural research has built over the years. It is the Government's stated policy that the gross national expenditure on research and experimental development (GNERD)

---

[2] The Science and Technology Act. Chapter 250 of the Laws of Kenya, revised edition 1980. Government Printer, Nairobi.

should attain a level of 1% GDP. The GDP in 1978–79 was K£1600 million, which suggests that the GNERD for that year should have been K£16 million. The country is therefore far from achieving the target of 1% GDP expenditure on R&D even if stock were taken of R&D funds allocated by parastatal organizations and the private sector.

The second functional level, management and tactics, involves promotion, sponsorship, and coordination of research. This is the responsibility of the advisory research committees (ARCs) catering for the various science groups (agricultural, medical, industrial, and natural sciences). Some of the functions of the ARCs are: (1) to advise on the details of research programs and projects required to implement the research priorities arising from the national science policy; (2) the concomitant budget requirements so arising; (3) to adivse on a quinquennial (or other suitable period related to the Development Plan) program for research and the estimates of the concomitant budget; and (4) to review, annually, the progress in research and to prepare for each Ministry its detailed program and expenditure for the following year.

The ARCs constitute the R&D circuit through which the technical ministries and executive level institutes are linked. The ARCs serve as forums for the establishment of research programs. ARCs are composed mainly of research scientists at the executive level, because the primary interest of the policymaking officers is taken care of at the science policymaking level (NCST). Their membership is so spread by discipline as to assure that the research programs have a full complement of research projects to solve the problems giving rise to the demands for research. They are also concerned with the quality and efficiency of the scientific research that comes under their review.

Table 2. Government expenditure (K£'000) in research and development (R&D) and scientific and technical services (STS) during the 1979–80 fiscal year.

Type of science	R&D		STS	
	Amount	% of total	Amount	% of total
Agriculture	7611	70.2	41074	43.5
Natural	1240	11.4	4503	4.8
Medical	1234	11.4	22812	24.2
Industrial	557	5.1	18263	19.4
Social	—	—	3312	3.5
Physical	—	—	548	0.6
Other	201	1.9	3870	4.1
Total	10844	100	94382	100

In the agricultural sector this responsibility lies with the Agricultural Sciences Advisory Research Committee (ASARC). The ARC is concerned with research activities related to agriculture, livestock development, forestry, wildlife, and water development. ASARC is required to work closely with the boards of management of the various research institutes, the management committees of the technical services of the Government, as well as the associations and unions of the private sector.

The third functional level, execution and operations, involves the carrying out of actual projects for scientific research, technical services, innovations, and education and training on institutional basis. These are the institutions that receive resources for R&D and are responsible for the implementation of the program of action and technical services.

This circuit is largely internal to the research establishments themselves. Representatives of NCST and ARCs are, by legislation, entitled to participate in the boards of management of the R&D institutions. In this way, the institutions are linked to both ASARC and NCST.

Criteria for Resource Allocation to Agricultural Research

Provisions of the Development Plan

In Kenya, the socioeconomic development policy is spelled out in 5-year development plans that have been in existence since independence. In these plans, the goals to be met are established. The activities of the public and private sectors are subsequently geared to meet the stated objectives.

Simultaneously, the National Council for Science and Technology analyzes the scientific and technological components of the development plan programs of action and publishes the National Science Policy for the plan period, out of which an assessment of the financial demand for research is undertaken. The financial requirements for R&D in the Government sector during the Fourth Development Plan (1979–83) for all sectors of the economy are shown in Table 3.

The theme of the Fourth Development Plan (1979–83) is "alleviation of poverty" through the provision of "basic needs" (food and nutrition, health, water, housing, and education). Because most social and economic problems such as poverty, malnutrition, disease, unemployment, and illiteracy are found in the rural areas, where about 85% of the population lives, the alleviation of these problems is

Table 3. Budgetary provisions (K£'000) to Government research institutions during the 1979–83 plan period.[a]

	1978–79	1979–80	1980–81	1981–82	1982–83
Recurrent	5812	6352	7573	8742	9947
Development	3805	4971	4651	4937	5093
Total	9617	11323	12232	13679	15040

[a] Source: Science and Technology for Development, a report of the National Council for Science and Technology, NCST Publication No. 4, May 1980.

foreseen to lie in the development of agriculture in order to lead to greater productive employment and income growth. Thus, the greater share of the R&D budget is devoted to agricultural research. The planned expenditure is shown in Table 4.

Commodity Research Institutions

Research in export commodities is mainly financed by commodity marketing boards with only token contribution by the Government. Thus, research in coffee is financed by the Coffee Board of Kenya and in tea by the Tea Board of Kenya. The Ministry of Agriculture has, however, considerable influence on the research policy regarding commodity research, mainly through participation in the boards and research management committees.

Commodity research institutions have a relatively higher level of expenditure per scientist. They are also better placed to attract and retain high calibre research scientists than similar institutions in the Government sector. They are also in a position to call upon the assistance of expertise from Government research institutions when such expertise is not available in their institutions.

The other type of commodity research institutions are those that deal with industrial crops that constitute the basis of local agro-based industries. Research in these commodities is financed jointly by Government and industry. For example, research in pyrethrum, sugarcane, and irrigation receive support from the Pyrethrum Board of Kenya, the Kenya Sugar Authority, and the National Irrigation Board, respectively. Similarly, the National Cereals and Produce Board provides grants to maize and wheat research.

Disbursements to Research Establishments

Government supported research stations prepare, annually, a 3-year forward budget and estimates of recurrent and development expenditure for the year in question. The submissions are aggregated at the ministry level for discussion with the Treasury. Simultaneously, requirements for additional personnel take place with the Directorate of Personnel Management in the Office of the President. Both the approved estimates of expenditure and personnel establishment are published in the Estimates of Recurrent and Development Expenditure in an aggregated form. Thereafter the responsibility for disbursement of resources (usually lower than what was bid for) lies with Directors of Research in various Ministry headquarters. No consultations take place with the directors of research stations regarding the allocation of the resources that have been availed.

Factors that influence the disbursement of resources include the provisions of the development plan and the science policy, initiation of new research programs, traditional practices whereby certain research stations are financed at a certain level irrespective of the research program content, pressure by the farming community and marketing boards, foreign aid donor supported projects, which have priority in the allocation of resources, and the influence of directors of research institutions.

Table 4. Agriculture research budget (K£'000) for Government research establishments during the 1979–83 plan period.

	1978–79	1979–80	1980–81	1981–82	1982–83	Total
Veterinary research	612	576	616	647	726	3177
Range research	404	396	372	432	417	2021
Animal production	734	714	833	732	773	3786
Crops research	2552	3034	3517	3795	4190	17088
Soils and seeds	297	230	236	285	310	1358
Economic research	280	660	780	800	700	3220
Joint services research	1583	1771	1946	2142	2354	9796
Total	6462	7381	8300	8833	9470	40446

Suggested Areas of Improvement

Decision-Making Process

The decision-making machinery that has been established in Kenya is considered adequate for resource allocation to agricultural research. Although the National Council for Science and Technology and the Agricultural Sciences Advisory Research Committees are both advisory, they are so constituted that their decisions have real authority and onus of implementation. The presence in the NCST of the Permanent Secretaries of the concerned technical ministries (Agriculture, Livestock Development, Natural Resources, and Water Development), the Permanent Secretary for Economic Planning and Development (socioeconomic development strategy), and the Permanent Secretary for the Treasury (the ultimate provider of resources) ensures that the interest of agricultural research is catered for. Similarly, ASARC's role of advising on the program of research and the concomitant budget required to implement the national science policy ensures a balanced allocation of resources among the various research programs.

Therefore, there appears to be no need to modify the existing decision and policymaking machinery. What is lacking at present is a solid data base on which to base research allocation in the future. To rectify this, the NCST and ASARC, with some financial assistance by IDRC, are undertaking a study on resource allocation at the institution, program, and project level. The study is designed to reveal resource allocation in relation to: (1) the agricultural commodity in terms of acreage, volume of production, monetary value, and nutritional value; (2) distribution among the various research agencies; (3) geographical and agroecological coverage; (4) personnel competence in terms of education and experience; (5) complementarity in terms of scientific disciplines; (6) adequacy of supporting staff and facilities; and (7) rationale on which the managers of agricultural research base the allocation of resources to various research programs and institutions.

It is expected that once the above study is completed, a better system of resource allocation will evolve.

Project Approval

Basically there are two sources from which demands for agricultural R&D emanate: (1) the need for development (from the farmer, extension service, or Government), which recognizes the need for research to provide answers to the technical problems inhibiting production; and (2) the research worker who envisages a breakthrough that will bring change to the understanding or technical control of his field and its application.

The first is now largely reflected in the Development Plan that, although not the basic source of detail, indicates the general plan and priorities accorded by the Government. Thus there are two influences at work in determining priorities. The first, and more traditional, is the upward demand for support of R&D based on the ideas and aspirations of research workers. The second, which is more recent, is the downward diffusion of policy based on the socioeconomic requirements of the country as stated in the Development Plan and the national science policy.

A framework for program approval has now been established. As noted earlier, NCST has the responsibility for the science policy while ASARC is concerned with the details of research programs. At lower levels, a system has been established that enables the translation of the policy framework to concrete projects and experiments. The system comprises:

(1) Provincial Research Advisory Committees comprising senior extension officers and farmer representatives in a particular province and research scientists undertaking research in the geographical area. The Committee is chaired by the Provincial Director of Agriculture. In this forum the extension service states the factors limiting production and research scientists develop research programs to solve the problems.

(2) Specialist Research Advisory Committees that deal with specific disciplines of countrywide concern. Thus, specialist committees exist for commodities such as maize, sugarcane, wheat, and pyrethrum, and specialist disciplines such as soil science, plant pathology, and entomology. Specialist committees comprise research scientists from other institutions and research scientists actually dealing with the particular commodity or discipline.

Project Costing

Project costing is an essential prerequisite for proper management of an agricultural R&D system. Without it, only very broad allocations of resources are possible, evaluation remains a vague activity, and measurements of efficiency and estimations of the cost-benefit of R&D work cannot be undertaken. At present, it is often the practice in Kenya to be satisfied with the simple statement that a research station, for example a commodity-oriented one, is carrying out research on that commodity. Even development plans allocate money to research stations merely identified by location, with no indication as to really what activities are intended. Annual reports, often years behind, produce descriptions of research and its results, the evaluation of which is often scientific and lacks other criteria. The efficacy and cost, both important factors, are not shown.

Research projects should therefore not be regarded as open-ended. Even where a follow-through or continuous activity over lengthy periods is a natural sequence to the establishment of a program, its various stages should be described in the form of completed or new projects. This is essential if account for the use of the national resources is to be sensibly displayed and a mark of progress is to be provided.

It is therefore suggested that the whole agricultural R&D scene must be accounted for and costed. Thus, any complete unit of R&D must be: (1) identified by a brief statement of its purpose; (2) costed in terms of estimated expenditure on manpower, material, and overheads; and (3) assessed in terms of duration.

Project identification and costing would provide the means by which the director of a research institution could display: (1) the use of his resources in a manner most meaningful to the higher levels of administration and policymaking (external use); (2) the actual activity and efficiency of his staff, which is a useful guide when considering their preferment (internal use); (3) the cost of research in relation to the value of the commodity concerned (economic use); and (4) the magnitude of the problem under research (scientific use).

Basic and Applied Research

There is considerable debate on the merits and demerits of basic research in a developing country such as Kenya. This controversy requires rationalization because it concerns allocation of scarce resources. The primary distinction between basic and applied research is that basic research produces knowledge and applied research produces know-how or technology.

Applied research is dependent, in the long run, on the results of basic research. New technologies, however, often make basic research feasible, after recognition of the need for better techniques or after acquisition of better data on which basic research is dependent. Therefore, both are, in their own right, dependent on the original ideas on what would be either interesting to know (basic) or useful to be able to do or produce (applied). The end product of basic research is manifest in the stimulation of further ideas. The end product of applied research is material production.

In Kenya the national science policy recognizes the need to support and allocate resources to basic research for the following reasons: (1) some applied research projects require basic research inputs to provide new avenues for further advancement; (2) the educational factor because, at postgraduate and higher levels, education becomes self-acquired through research, which could well be basic in nature; (3) the need to maintain and increase scientific excellence in the scientific establishments in the country; (4) the need to obviate the masquerading of basic research as applied research when bidding for resources; and (5) the factor of employment, particularly in respect of highly specialized scientists whose loss (by degrading their work or by their emigration) would, in the long run, amount to brain drain.

The initiative for basic research should, in the main, lie with individual scientists and research establishments. There is need to establish criteria against which resources would be allocated to basic research. The suggested criteria should include: (1) the scientific merit and efficiency of the institution and research scientists proposing the project; (2) the relevance of the project in scientific, economic, social, environmental, and political terms; (3) the priority of the project in terms of the national socioeconomic policy; (4) the predictability of results; (5) duration of the project; and (6) the cost of the project.

To the above criteria should, ultimately, be added considerations of the ratio of the total cost of R&D of all projects relevant to a given commodity in relation to its value. The criteria should be weighted according to the nature of the research project. Because applied research is heavy in socioeconomic terms, basic research should be heavy on personnel merit and scientific relevance.

It is suggested that resources availed to basic research in agriculture should be in the order of 5% of the Gross National Expenditure on Research and Experimental Development devoted to agriculture.

Resource Allocation to Agricultural Research in Bangladesh

Ekramul Ahsan[1]

Agriculture plays an important role in Bangladesh's economy, accounting for 55% of the gross national product and 85% of all employment. It generates 80% of the country's export, with jute,tea, and hides and skins the leading foreign exchange earners. About 22.5 million acres (9 million hectares) are under cultivation with an average cropping intensity of about 147%. Agriculture is characterized by traditional farming, numerous small farms, and low capital investment.

Agricultural research in Bangladesh dates back to 1880 when a Division of Agriculture was established under the Department of Land Records in Bangal on the recommendation of Finance Commission. Later in 1906, the Department of Agriculture was established with separate identity and status. Subsequently, an agricultural research laboratory was established in 1908 at Tejgoan, Dacca, to serve the provinces of Bangal and Assam. This laboratory persisted through a series of reorganizations during the three decades since partition of the subcontinent in 1947 and evolved into the Agricultural Research Institute of East Pakistan.

Agricultural research was seriously disrupted in 1962, when the experiment station lands of the agricultural research institute were taken for building the second capital of Pakistan. Throughout the 1960s, when most of the Asian countries were modernizing agricultural production through the adaptation and use of improved technology, Bangladesh lost its institutional base to participate in this modernization process.

The stagnated status of agricultural research after 1962, resulting from the disarray of the Agricultural Research Institute and the difficulties in carrying out meaningful research under the constraints of government administrative procedures, tended to accelerate actions to set up separate, and in many cases autonomous, research institutes or centres. The following institutions for agricultural research are now found in the country: Bangladesh Agricultural Research Council (BARC); Bangladesh Agricultural Research Institute (BARI); Bangladesh Rice Research Institute (BRRI); Bangladesh Jute Research Institute (BJRI); Sugarcane Research Institute (SRI); Institute of Nuclear Agriculture (INA); Tea Research Institute (TRI); Livestock Research Centre; Fisheries Research Institute; and Forest Research Institute.

Agricultural Research System

Agricultural research, a vital instrument of progress in the process of developement in Bangladesh, did not receive sufficient attention in the past. Research stations and facilities were underdeveloped, staff training lagged, scientists were isolated from relevant national and international research programs, the responsibility of agricultural research lay within several different ministries and agencies, and coordination was weak or absent. Moreover, the failure was due to inadequate research backing for agricultural development efforts. Since 1973, however, it has been increasingly recognized that the national research system must be further strengthened or improved to provide new means of improving food and agricultural production.

Research efforts in Bangladesh are highly fragmented with at least nine ministries involved in administering research on various aspects of agriculture (Table 1). The research capabilities of the research institutions have also been developed to serve specific commodities or specialized sectors of agriculture because of the influence of special interest groups. A continuing and increasingly critical limitation to the development of a well-integrated national agricultural research system is the dispersal of research institutes and centres, with their substantial autonomy, throughout nine different ministries

[1]Member Director, Agricultural Research Council, Farm Gate, Airport Road, GPO Box 3041, Dacca, Bangladesh.

129

Table 1. Institutions conducting agricultural research in Bangladesh.

Ministry of Agriculture and Forests
    Bangladesh Agricultural Research Council
    Bangladesh Agricultural Research Institute
    Bangladesh Rice Research Institute
    Forest Research Institute

Ministry of Livestock and Fisheries
    Livestock Research Centre
    Fisheries Research Institute

Ministry of Education
    Bangladesh Agricultural University
    Bangladesh University of Engineering and Technology
    Dacca University
    Chittagong University
    Rajshahi University

Ministry of Science and Technology
    Bangladesh Jute Research Institute
    Bangladesh Council of Scientific and Industrial Research
        Regional Laboratories at Dacca, Chittagong, and Rajshahi
    Bangladesh Atomic Energy Commission
        Irradiation and Pest Control Research Centre
        Institute of Nuclear Agriculture

Ministry of Planning
    Bangladesh Institute Development Studies

Ministry of Industries
    Sugar Mills Corporation
    Sugarcane Research Institute

Ministry of Commerce
    Tea Board
    Bangladesh Tea Research Institute

Ministry of Local Government and Rural Development
    Cooperatives and Rural Development
    Bangladesh Academy of Rural Development at Comilla and Bogro

Ministry of Power Water Resources and Flood Control
    Water Development Board
    Land and Water Use Department

of the government. The duplication of facilities and programs is a serious constraint to building a strong and effective national research system.

In an effort to coordinate the national research program, the Bangladesh government established the Bangladesh Agricultural Research Council (BARC) in 1973 to provide the guidance and integration of research on all aspects of agriculture: crops, livestock, soils, water, crop protection, agricultural engineering, forestry, fisheries, and economic and social sciences. But BARC had influence only over the research institutes financed by the Bangladesh Agricultural Research Council. Research programs continued in isolation and communication among workers in various institutes was almost totally lacking.

The council could not perform the functions needed for better planning, coordination, and consolidation of research. The existing agricultural research systems in Bangladesh have been studied, reviewed, and evaluated by a number of experts and teams. All of their reports have emphasized the need for consolidation of all agricultural research undertaken by the various institutes. This effort, under the Bangladesh Agricultural Research Council, would help align research with national needs and priorities, speed up the generation of dependable research results and their application, and increase efficiency by reducing wasteful duplication. The Ministry of Agriculture and Forests has recently made a significant effort to strengthen research coordination by instructing all research institutes under its control (BARI, BRRI, and FRI) to process their research projects through BARI and obtain the council's approval before submitting them to the government. Other ministries involved in agriculture research have not yet taken similar steps.

Organization

The major constraint to building a strong, effective, and efficient research system in Bangladesh is the fragmentation of research effort under various ministries (Table 1). The research institutes have tended to function independently, having no well-defined role or responsibility as a component of an integrated national research system. Some of these institutes, with substantial technical and financial assistance from external sources, are almost self-sufficient and lack a sense of urgency to cooperate with other research organizations. Incentive is lacking among these institutes to respond to the initiatives of BARC and to address factors of broad national concern in agricultural research and development. The following outline of the establishment and function of the individual research institutes reflects the nature and diversification of agricultural research in the country. However, efforts are under way to coordinate the national research systems through BARC to strengthen the research capabilities of individual institutions by a planned and integrated program of resource allocation.

Bangladesh Rice Research Institute

The takeover of experimental land at Dacca for the construction of the second capital hampered the activities of the East Pakistan Agricultural Research Institute. Since the late 1960s, steps have been taken to establish research capabilities for some of the important commodities and components of agriculture.

One of the first efforts was directed to improving rice production technology with assistance from the Ford Foundation and the International Rice Research Institute. The rice research program was initially carried out at Saver Dairy Farm and at Joydebpur through the implementation of the East Pakistan Accelerated Rice Research Scheme. Although this was a program of the Agricultural Research Institute under the Agriculture Department, it was the first attempt at a multidisciplinary approach to monocrop research in the country.

The Rice Research Institute was established in 1970 with the status of a semiautonomous organization for acclerating rice research. It was administered by a board with authority to formulate and execute its policies and strategies within a framework of general policy directions from the government.

After independence, the Bangladesh Rice Research Institute (BRRI) was granted the status of a fully autonomous institute to carry out research for the improvement of cultivation and the production and development of improved rice varieties and other improved technologies.

Bangladesh Agricultural Research Institute

The Bangladesh Agricultural Research Institute (BARI), the premier agricultural research organization in the country, was a government institution under the Directorate of Agriculture until 1975 and faced financial as well as operational constraints. It became increasingly clear that the lack of flexibility in administrative matters, financing, staffing, procurement, and general operations was seriously hampering the accelerated development of an effective multidisciplinary research capability. After a series of intensive studies to reorganize and strengthen the institute, in January 1976 the institute was shifted to a new site at Joydebpur and given autonomous status with essential operational flexibility. Within a short time, BARI became the largest reseach institute in Bangladesh and began to foster broad commodity research as well as research in noncommodity areas such as crop protection, agronomy, soil fertility, testing and conservation, agricultural engineering, and agricultural economics.

Bangladesh Jute Research Institute

The Bangladesh Jute Research Institute (BJRI) was first established in 1951 and was for several years under the Ministry of Agriculture. Subsequently, to provide flexibility and freedom from government administrative constraints, it was set up under the Ministry of Jute in 1973 and given semi-autonomous status. The Jute Research Institute has made substantial progress in improving facilities, upgrading research personnel, and evolving a more relevant and effective research program. Recently, it has been transferred to the Science and Technology Division of the Cabinet Secretariate.

Sugarcane Research Institute

The Sugarcane Research Institute (SRI) was originally under the Ministry of Agriculture but was subsequently transferred to the Ministry of Industries and has operated under the Bangladesh Sugar and Food Industries Corporation (BSFIC) since 1973. This transfer was made with the expectation that the corporation would make substantial investments to improve the research capabilities of the institute. However, a number of problems exist that are inherent in the present status of SRI and its relationship to the corporation.

Institute of Nuclear Agriculture

The Institute of Nuclear Agriculture (INA) was established in 1972 under the Atomic Energy Centre

Table 2. Resource allocation to agricultural research in Bangladesh, 1975–76 to 1979–80.

	Total allocation		Domestic sources			External resources			External/ Domestic
	Million Tk		Million Tk		% of total	Million Tk		% of total	
1975–76	103.5[a]	(103.5)[b]	63.9	(63.9)	61.8 (61.8)	39.6	(39.6)	38.2 (38.2)	0.62 (0.62)
1976–77	99.8	(89.7)	78.7	(70.7)	78.9 (78.8)	21.1	(19.0)	21.1 (21.2)	0.27 (0.27)
1977–78	164.5	(131.7)	97.7	(78.2)	59.4 (59.4)	66.8	(53.5)	40.6 (40.6)	0.68 (0.52)
1978–79	192.6	(134.9)	125.4	(87.9)	65.1 (65.1)	67.2	(47.1)	34.9 (34.9)	0.54 (0.40)
1979–80	379.9	(242.0)	233.5	(148.7)	61.5 (61.5)	146.4	(93.2)	38.1 (38.5)	0.63 (0.26)
Total	940.3	(701.8)	599.3	(449.4)	63.7 (64.0)	341.1	(252.4)	36.3 (36.0)	0.57 (0.56)
Growth rate	26.00	(16.99)	25.91	(16.89)	— —	26.10	(17.12)	— —	— —

[a] Expressed at current price.
[b] Expressed at constant price (1975–76).

to foster the application of nuclear technology to agricultural problems. The program of the institute has, however, been moved toward traditional plant breeding and genetics research, evaluation of nutritional qualities of cereals and legumes, and other related areas. The research at INA on rice, jute, and other commodities and production factors is not well coordinated with similar work at other organizations.

Tea Research Institute

The Tea Research Institute operates under the Tea Board of the Ministry of Commerce. It has very little linkage with other research programs and institutes and is isolated physically as well as in terms of program coordination.

Forest Research Institute

The Forest Research Institute, which is engaged in research to develop forest resources, is under the Ministry of Agriculture and is administered under the rules and regulations of the government.

Resource Allocation to Agricultural Research

To study the resource allocation system for agricultural research, information was collected to determine the trend in the allocation of resources to commodity and noncommodity research in the various research organizations in Bangladesh. The total investment in agricultural research in the various organizations was broken down to capital items, recurring costs, and operational costs. The influence of external resources was also examined. Most of the descriptions refer to the period between 1972 and 1980 and are based on two surveys: one undertaken in 1978 by BARC; the other jointly by BARC and ADC in early 1981. A comprehensive survey of resource allocation in Bangladesh agricultural research was initiated recently with assistance from IDRC, but it is still too early to report the results.

Table 3. Resource allocation to agricultural research per capita and as a percentage of GDP (1975–76 to 1979–80).

	Resources per capita (million Tk)		Percentage of GDP
1975–76	12.9[a]	(12.9)[b]	0.17[b]
1976–77	12.2	(11.0)	0.16
1977–78	19.7	(15.8)	0.23
1978–79	22.5	(15.8)	0.24
1979–80	43.5	(27.7)	0.38

[a] Expressed at current price.
[b] Expressed at constant price (1975–76).

Investment in agricultural research is very low in Bangladesh, but the returns from resource allocation in research give good results. In recent years, Bangladesh has increased her investment in research to 0.1% of GDP. This is still very low compared with many developing countries and is grossly inadequate to meet the challenge of doubling food production. Bangladesh, with its potential human and natural resources and present challenges, provides a vast opportunity for a high payoff from research. There is no definite criterion to precisely specify the optimal level of expenditure on scientific research in agriculture; however, an increase in allocation to 0.7% of GDP is desirable to achieve the projected development goals.

The following criteria were mentioned in the National Agricultural Research Plan of Bangladesh:[2] (1) contribution of agriculture to GDP (56%); (2) contribution of agriculture to foreign exchange earnings (95%); (3) contribution of agriculture to employment generation (85%); (4) unlimited scope for futher development in agriculture; and (5) criteria to be selected from experience of other countries. Based on these criteria, it is suggested that the allocation for agricultural research be about 75% of the total allocation for research. It is necessary to ensure that such allocations be made in a manner that removes sectoral imbalance within agriculture.

It is interesting, but more important, to note that until recently, more than 60% of the allocation on research was provided for capital investment in physical infrastructure. This trend should be reversed so that proportionately more is spent on operational purposes. Manpower development should also receive higher priority to increase the efficiency of research.

Total Allocation to Agricultural Research

An April 1981 study revealed that in the most recent financial year, 42 million taka have been allocated to operational funding of agricultural research, whereas the requirement was estimated to be 62 million taka, i.e., there is a 36% shortfall in operational funds.

Table 2 shows the resources allocated to agricultural research in the past 5 years (1975–80) at current prices. Allocations to agricultural research at current prices have been growing at a rate of 26% per annum. The ratio of total domestic resources to external resources for the 5-year period is 0.57 at current prices, which shows that domestic resources

____

[2] BARC. Strengthening the Bangladesh agricultural research system: report of the joint research review team. Dacca, April 1979.

Table 4. Total capital expenditure (million Tk) at major institutes and percentage distribution of total allocation (1975–76 to 1979–80).

	1975–76		1976–77		1977–78		1978–79		1979–80		Total capital expenditure (5 years)	Percentage of total allocation within institution	Percentage of total capital	Growth rate
	Total	%	Total	%	Total	%	Total	%	Total	%				
BARI	NA	—	11.41	47.3	23.25	55.4	49.35	64.0	81.21	62.8	165.22	60.6	46.51	49.06
BRRI	12.01	68.4	18.17	74.9	14.08	63.2	16.89	59.6	19.91	49.9	81.05	61.3	22.82	28.48
BJRI	2.70	77.7	5.63	82.8	3.96	43.3	6.94	68.0	5.71	51.0	24.94	61.4	7.02	14.98
SRI	1.96	58.0	0.89	31.8	2.02	54.8	7.65	74.0	10.42	76.0	22.94	67.9	6.46	33.42
TRI	2.36	54.8	1.48	44.0	2.06	45.4	1.98	32.9	0.71	13.9	8.59	36.8	2.42	—
INA	6.78	71.8	4.59	57.3	1.40	31.9	3.75	44.1	0.17	4.3	16.69	48.6	4.70	—
BARC	0.49	31.0	0.32	12.1	0.15	7.1	0.35	7.0	5.59	52.9	6.90	31.71	1.94	48.69
FRI	2.32	47.8	5.41	63.1	6.43	57.2	6.42	52.6	8.29	54.5	28.87	55.4	8.13	25.64
Total	28.62	8.06	47.90	13.48	53.35	15.02	93.33	26.27	132.01	37.16	355.21	—	100.00	31.56

Table 5. Total net operational cost (million Tk) of different research institutes and percentage distribution of total allocation.

	1975–76		1976–77		1977–78		1978–79		1979–80		Net operational cost (5 years)	Percentage of total allocation within institute	Percentage of operational cost	Growth rate
	Total	%	Total	%	Total	%	Total	%	Total	%				
BARC	0.54	36.4	0.80	30.5	0.81	39.0	1.99	39.8	2.75	26.0	6.89	31.6	6.84	32.56
BARI	—	—	4.41	18.3	4.78	11.4	6.98	9.0	12.76	9.9	28.93	10.6	28.71	26.62
BJRI	0.26	7.6	0.35	5.1	0.68	7.4	0.69	6.8	0.72	6.6	2.70	6.7	2.68	20.37
BRRI	2.20	12.5	3.23	13.3	4.18	18.8	5.11	18.0	11.99	30.1	26.71	20.2	26.51	33.91
FRI	0.99	20.4	1.33	15.5	1.56	13.8	1.77	14.5	2.56	16.8	8.21	15.8	8.15	19.00
INA	1.89	20.0	2.55	31.8	1.76	40.1	2.99	35.2	1.70	42.6	10.89	31.7	10.81	—
SRI	0.91	26.9	1.33	47.4	0.86	23.2	1.65	16.0	1.89	13.9	6.64	19.6	6.59	14.62
TRI	1.31	30.5	1.20	35.6	1.47	32.4	2.71	45.1	3.10	60.6	9.79	42.0	9.71	17.23
NOC[a]	8.10	7.31	15.2	15.08	16.1	15.98	23.89	23.71	37.47	37.19	100.76	—	100	30.36

[a] NOC = Net operational cost.

have been almost twice the external resources available for agricultural research. However, the year-to-year picture shows that the level of external resources has varied. Although a specific trend in the ratio of domestic resource allocation to external research allocation cannot be derived, an estimate of the growth rate of allocations for both domestic resources and external resources for agricultural research indicates a proportionately higher rate of investments of foreign resources in agricultural research in Bangladesh. The figures in parentheses in Table 2 indicate the allocations in constant prices and provide a similar picture.

Table 3 shows that resource allocation for agricultural research as a percentage of gross domestic product (GDP) at constant 1975–76 prices has increased from 0.17% of GDP in 1975–76 to 0.38% of GDP in 1979–80. This indicates a positive trend in the allocation of resources as a percentage of gross domestic product over time although it is still below the level of 0.7% of GDP for agricultural research suggested in the National Agricultural Research Plan of Bangladesh.

In Bangladesh, it has been estimated that the internal rate of return from agricultural research is about 30%. But, in spite of the prospect of this high rate of return to investment, allocations for agricultural research in Bangladesh are low both in relative and absolute terms. However, the consistent increase in the past several years is a healthy sign and a desirable policy.

Resource Allocation to Research Organizations

Resources have been allocated in favour of capital investments and less attention has been paid to adequate allocations for operational costs (Tables 4 and 5). It is logical and desirable that, in the initial development of a research organization, high capital investment is required to develop the basic facilities and infrastructure for research. But, in Bangladesh it is perhaps time to increase allocations to meet operational costs to make the best use of capital investments and research facilities.

A breakdown of resource allocations to the major agricultural research institutes reveals that some institutes have received reasonably high investments for both capital and operational costs, but other organizations have shown a very unfavourable trend.

Institutional characteristics reflect the funding position of the institute within the different ministries and the funding given to these ministries. The length of establishment of an organization also has an influence. Some of the older institutes such as TRI (established 1958), INA (1964), and JRI (1954) have largely completed their program of major capital investment for physical facilities, therefore the

Table 6. ADP allocation for agricultural research.

	1975–76	1976–77	1977–78	1978–79	1979–80	Total	Average
BARI	29.29	35.57	55.70	67.10	109.22	296.68	59.34
	(28.05)[a]	(31.73)	(35.57)	(39.94)	(46.64)	(38.31)	
BRRI	25.51	27.15	30.10	26.60	27.46	136.82	27.36
	(24.42)	(24.36)	(19.22)	(15.86)	(11.73)	(17.66)	
BJRI	20.69	24.45	45.34	52.00	43.20	185.68	37.14
	(19.82)	(21.94)	(28.94)	(31.00)	(18.45)	(23.97)	
INA	12.50	5.40	5.00	2.50	6.00	31.4	6.28
	(11.97)	(4.85)	(3.19)	(1.49)	(2.56)	(4.05)	
FRI	6.46	5.55	8.10	8.00	18.50	46.61	9.32
	(6.19)	(4.98)	(5.17)	(4.77)	(7.90)	(6.02)	
SRI	6.50	4.20	2.50	3.50	2.00	18.7	3.74
	(6.23)	(3.77)	(1.60)	(2.09)	(0.85)	(2.41)	
BARC	1.80	4.50	3.60	4.00	21.20	35.10	7.02
	(1.72)	(4.04)	(2.30)	(2.39)	(9.05)	(4.53)	
Veterinary Res. Institute	0.91	2.69	3.56	1.50	2.95	11.61	2.32
	(0.87)	(2.41)	(2.28)	(0.89)	(1.26)	(1.50)	
Land and water use research	0.65	1.95	2.45	2.50	3.40	10.95	2.19
	(0.62)	(1.75)	(1.57)	(1.49)	(1.45)	(1.41)	
Agricultural marketing research	0.10	0.20	0.25	0.15	0.26	0.96	0.19
	(0.09)	(0.18)	(0.16)	(0.09)	(0.11)	(0.12)	
Total	104.41	111.46	156.6	167.85	234.19	774.51	

[a] Percentages given in parentheses.

134

Table 7. Gap between ADP allocation (ADP) and net available fund (NAF) (1975–76 to 1979–80) (figures in million Tk).

	1975–76			1976–77			1977–78			1978–79			1979–80		
	ADP	NAF	Gap	ADP	NAF	Gap	ADP	NAF	Gap	ADP	NAF	Gap	ADP	NAF	Gap
BARI	29.29	NA	NA	35.37	24.12	11.25	55.7	41.96	13.74	67.00	77.15	-10.15	109.22	129.23	-20.01
BRRI	25.51	17.55	7.96	27.15	24.26	2.89	30.1	22.27	7.83	26.60	28.33	-1.73	27.46	39.87	-12.41
BJRI	20.69	3.47	17.22	24.45	6.80	17.65	45.34	9.14	36.2	52.00	10.21	41.79	43.20	11.00	32.20
INA	12.50	9.44	3.06	5.40	8.01	-2.61	5.00	4.38	0.62	2.50	8.49	-5.99	6.00	4.00	2.00
FRI	6.46	4.86	1.60	5.55	8.58	-3.03	8.10	11.24	-3.14	8.00	12.20	-4.20	18.50	15.19	3.31
SRI	6.50	3.38	3.12	4.20	2.80	1.40	2.50	3.69	-1.19	3.50	10.33	-6.83	2.00	13.59	-11.59
BARC	1.80	1.48	0.32	4.50	2.62	1.88	3.60	2.08	1.52	4.00	5.00	-1.00	21.20	10.58	10.62
Total	102.75	40.18	62.57	106.62	77.19	294.21	150.34	94.76	55.58	163.60	151.71	11.89	227.58	223.46	4.12

proportion of capital investment out of the total allocation is falling quite steadily. On the other hand, BARI (established as a separate entity in 1976) is in the midst of implementing a large-scale program of developing physical infrastructure and the allocation of capital investment is still increasing in comparision with recurrent expenditures and more particularly with respect to operational costs (Tables 4 and 5).

The experience of many agricultural research organizations in Bangladesh indicates that the amount available for operational costs has been inadequate to undertake meaningful research of national importance and to make the best use of the equipment and facilities. Within this unfavourable situation, certain institutes, namely, BRRI, TRI, and SRI, are in a relatively better position because they receive operational funds from sources outside government budgetary allocations.

Total ADP Allocation to Agriculture

Budgeting procedures differ depending on the research institution and source of funds. The Annual Development Programme (ADP) at the national level is allocated to different research institutions by the Planning Commission. The Planning Commission, after reviewing the total resources available, provides the ADP allocation according to the criteria for annual development in the agricultural sector. Such criteria are, however, not always well-defined. Of the total allocation to the agricultural sector, allocations to individual research institutions are made based on the priority of their programs. Once again, specific criteria are not always clear.

Table 6 shows the ADP allocation to different research institutions during 1975–80. It reveals that the greatest allocation was made to BARI. Up-to-date information on the allocation to fisheries and livestock research could not be incorporated, but according to an earlier survey, these are the two areas that are historically neglected in Bangladesh (fisheries 0.81% and livestock 1.03% of total allocation in 1976).

Table 7 reveals the gap between ADP allocations and net available fund during 1975–80. In most of the institutions, the net available fund has always been lower than the ADP allocation. However, the situation has improved, particularly with respect to BARI, BRRI, and SRI.

External Financing of Agricultural Research

External resources are available in the form of grants, donations (e.g., Ford Foundation, IDRC, USAID), and loans (e.g., IDA, ADD). An estimate reveals that 16% of the assistance is in the form of loans. Most of these external resources are used in capital investments and technical assistance through expatriate services and very little donor support is available to supplement the operational fund of the research programs. It is perhaps more desirable for donors to supplement operational cost to reap the fullest benefit from their investments in capital items. Another important area that needs more emphasis from external assistance is increased involvement in manpower development programs for research scientists.

I wish to express my gratitude to Dr. Kazi M. Badruddoza, Executive Vice-Chairman, Bangladesh Agricultural Research Council, for kindly nominating me to work on this study and for his encouragement. Sincere thanks are extended to Mr. P.C. Munasinghe of IDRC for helping to organize the study and for time to time consultation. The hard work done by Mr. Mufakhkharul Islam, Senior Scientific Officer, BARC, in organizing the survey and in preparation of the report is highly appreciated. The painstaking job of typing the paper by Mr. Golam Rabbani of BARC is acknowledged with thanks.

A Preliminary Attempt to Evaluate the Agricultural Research System in Brazil

**Maria Aparecida Sanches da Fonseca and
José Roberto Mendonça de Barros[1]**

The Brazilian economy has felt the impact of the reorganization of the world's economic system. With an increase of almost 1000% in the average price of crude oil between 1973 and 1980, Brazil has faced a significant increase in its oil import bills and, consequently, in the price of many industrial products. In 1973, 12.4% of the country's total import costs were accounted for by oil; in 1980 the figure reached 40%. Despite an increase in exports, the balance of payments deficit continues to increase.

There is a need for a major effort to pursue a balance of payments equilibrium. This is not an easy task because this goal must also: avoid a decline in the employment level; control inflation; and meet domestic supply and energy needs. In view of these problems, the agricultural sector has been asked to play a decisive role with respect to: domestic food supply; an increase in exports; and a reduction of food and fuel imports.

Most of all, the agricultural sector is expected to supply food in sufficient quantities and at relatively low prices. In fact, the government policy of helping to keep real wages from falling will not succeed unless the agricultural sector is able to provide food at prices that do not cause the cost of living to climb. On the other hand, agriculture is also expected to be a source of foreign exchange to maintain a trade balance. With respect to this aspect, the agricultural sector, through the role it plays in supplying food and raw materials to the industrial sector, must play a double role: supply foreign exchange to increase the country's import capacity; and avoid the import of food items and thus save foreign exchange. Since 1973, agriculture has also played an increasingly important role in energy production. In reality, the energy programs developed so far assign a preponderant role to agriculture for the production of biomass.

Good agricultural performance depends upon a number of factors, including availability of technology and a constant flow of knowledge. There is a consensus in the country about the need to generate and diffuse agricultural technology. Satisfactory performance of the agricultural sector will depend, to a very significant extent, on the process of technological change. On the other hand, the characteristics of this technological change depend on guidelines set by institutions responsible for both the generation and diffusion of technology. Agriculture's contribution to economic development depends on: (1) the policy chosen toward research and technical (extension) assistance; (2) the role played by both the public and private sectors in the generation and diffusion of technology; (3) the amount of resources allocated to these activities; (4) the balance between basic and applied research; (5) the distribution of resources among agricultural products; and (6) the supply of agricultural inputs (Silva et al. 1979).

The purpose of this paper is to analyze the research system set up at the federal level and in the state of São Paulo[2] and to assess the financial resources that have been made available to agricultural research. An attempt is also made to establish the relationships between the total resources available for agricultural research and the country's research and development (R&D) program, GNP, resources allocated to technical assistance, and international investment patterns in R&D. After the research organization structure and resource availability are analyzed, the resource allocation criteria are reviewed. The explicit resource allocation criteria are

---

[1]Centro de Fertilizantes, Instituto de Pesquisas Tecnológicas, Caixa Postal 7141, São Paulo, Brazil.

[2] The state of São Paulo, as opposed to other states in the country, has been investing in agricultural research since the nineteenth century.

considered on the basis of documents issued by research agencies and on interviews conducted with scientists. A survey of scientific publications related to agriculture in Brazil is also analyzed. Finally, the decision criteria that apply to the Brazilian research system, their limitations, and their implications are discussed.

Institutional Structure in Brazil

Policy for Science and Technology

For a better understanding of the institutional structure for agricultural research in Brazil, it is necessary to review Brazilian policy for science and technology because the first should be in accord with the second. When defining a policy for science and technology, one must try to use it as a means of reaching the major objectives of Brazilian society. In the economic arena, this means the country should be equipped to produce technology as well as consumer goods.

A policy for scientific and technological development has been included in the Basic Plan for Scientific and Technological Development (PBDCT), at least in terms of a theoretical program that is in accord with the broad objectives of the National Development Plan (PND).

The objectives of the policy for scientific and technological development vary from one plan to another, but the need to alter Brazil's dependence on foreign technology is recognized in all plans. According to the priority items found in the Third PND, which is now in force, energy, agriculture, social development, and the search for greater scientific capacity and a reduction in technological dependence constitute major policy guidelines addressed to: (1) reduce the country's need to import energy inputs; (2) augment the country's capacity to adequately select technologies among existing options; (3) promote effective absorption and adaptation of technologies; and (4) generate its own solutions in response to regional diversities.

The PBDCT plan is defined by the National Council for Scientific and Technological Development, which has the power to allocate government budget funds and several other sources of revenue to the financing of scientific and technological activities. Additional resources may be derived directly and indirectly from the federal administration, as well as from private companies. All these additional resources are channelled through special federal agencies that also handle funds from international sources.

The structure and operation of the country's agricultural system is not isolated but is inserted in a broader context that provides basic operational guidelines.

Institutions of Agricultural Research

The agricultural research institutions in Brazil may be linked with either the federal or state governments or with the private sector. Within the federal government, the research institutions can be divided into those under EMBRAPA and those that fall outside the EMBRAPA structure.

Brazilian Agricultural Research Company (EMBRAPA)

To accelerate and consolidate its agricultural research program, the federal government changed its specialized operational structure and created EMBRAPA in 1973 to replace the National Department for Agricultural Research under the Ministry of Agriculture. The newly created company assumed its research activities in 1974. Both direct and indirect methods of operation were established by EMBRAPA to meet its objectives.

Direct action is taken through the national centres and the executive research units at the state level (UEPAE). The national centres, located in various parts of the country, are used to generate technology through interdisciplinary effort and cover only a certain number of agricultural products of key national interest. In addition to these national centres there are: The National Center for Genetic Resources; The Agricultural Research Center for the Humid Tropics; The Agriculture Research Center for the Semi-Arid Tropics; and The Agriculture Research Center for the Cerrado Region.

These UEPAEs face the task of adapting technology developed by the national centres as well as generating technology for local products. In cooperation with the centres, they are responsible for generating technology for products considered important at the national level.

Indirect action is exerted mainly by the state agricultural research companies that obtain financial support from EMBRAPA and from the states themselves. These companies follow programs and norms imposed by EMBRAPA, which monitors and evaluates their activities. There are also some integrated programs resulting from special agreements between the Ministry of Agriculture and the state governments. These programs are under the supervision of EMBRAPA and the Agricultural Secretariat.

EMBRAPA also influences other special groups that, at the national level, carry out research programs of great importance: The National Service for Basic Seeds; The National Service for Soil Survey and Conservation; and The National Program for Product Processing Technology.

Federal Government Institutions not in EMBRAPA Structure

The Cocoa Research Center, The Brazilian Institute of Forestry Development, The Brazilian Coffee Institute, The Sugarcane Center (PLANALSU-CAR), Universities, and Regional Development Superintendencies are not included under EMBRAPA. Finally, there are the state research organizations. Within this group, São Paulo's research system stands out. It dates back to the nineteenth century and comprises nine agricultural research institutes: Forestry, Fisheries, Plant Science, Agricultural Economics, Geology, Agronomy, Biological, Food Technology, and Animal Science.

The participation of cooperatives, private companies, and universities in the development of agricultural research is increasing.

Resource Endowments for Agricultural Research

In spite of the importance of agriculture for Brazil, this sector did not receive significant resources until the beginning of the seventies, when there seems to have been recognition of the need for agricultural research.

The national resources allocated to agricultural research and to science and technology as a whole are available only for five years (Table 1). The percentage allocated by PBDCT to agricultural research increased from 10.8% to 15.3% during 1973–77.

EMBRAPA's resources (in U.S. dollars) were: 1973, 2.7 million; 1974, 33.4 million; 1975, 66.8 million; 1976, 93.1 million; 1977, 102.9 million; 1978, 118.7 million; 1979, 151.5 million; and 1980, 142.3 million. These figures do not include the resources derived from state governments in the case of the state agencies.

Table 1. Basic plan for scientific and technological development in Brazil, 1973–77 (figures in thousands of U.S.$). Source PBDCT; conversion rate Cr$52.699/U.S.$1.

	Funds for agricultural research	Total research funds	% total funds to agriculture research
1973	56 537	521 508	10.8
1974	70 343	614 926	11.4
1975	141 179	1 111 254	12.7
1976	175 297	1 241 390	14.1
1977	204 710	1 339 987	15.3

Silva et al. (1980a,b) estimated the expenditures in Brazilian agricultural research and technical assistance during the period 1974–78 (Table 2) and showed that at the national level they increased for both research and technical assistance. In the state of São Paulo, which traditionally allocates resources for research and technical assistance, these investments have been reduced.

The distribution of the EMBRAPA's resources among research programs is shown in Table 3. The table does not indicate the amount received by different regions for a given program, but it is known that the amount varies from region to region. A comparison between the amount destined to food products and export products does not show great differences. However, it is important to remember that this type of comparison cannot be made by considering only EMBRAPA's resources because EMBRAPA does not conduct research in products such as coffee, sugarcane, and cocoa, which are important foreign exchange earners. These products have their own research institutes. The figures for EMBRAPA also do not show the amount and the distribution per crop of the resources committed to research by the states that do not belong to the EMBRAPA system.

EMBRAPA's research staff has increased from 872 to 1553 in the period 1974–80. The National Agricultural Research System (EMBRAPA, State Agricultural Research Corporations, and Integrated Programs) employs 2935 researchers out of a total of 14 200 employees (Table 4).

Of the total number of researchers, 1684 have, since 1974, started graduate courses (both Master and Ph.D. programs) at universities in Brazil and abroad (Table 5). More than 500 have already returned to their research activities at EMBRAPA. EMBRAPA's postgraduate project, from which several institutions have already profited, is one of the greatest efforts ever made in such a short period by any institution in the world.

Although complete information is lacking about the activities of all of EMBRAPA's researchers, Table 6 gives an indication of their distribution by programs.

Comparison Between Agricultural Research Expenditures and Other Variables

The total expenditures on agricultural research activities in Brazil increased by nearly 200% during 1974–1978. The First and Second PBDCT show investments of 56.5 million dollars in 1973 and 204.7 million dollars in 1977 (constant price). Comparing these values with the total programed investment in the PBDCT, the percentage of expenditures

Table 2. Investments in agricultural research and technical assistance in Brazil and São Paulo, 1974–78 (figures in thousands of U.S.$ at constant prices). Source: Silva et al. 1980b; conversion rate Cr$52.699/U.S.$1.

	Brazil			São Paulo		
	Research	Technical assistance	R/TA	Research	Technical assistance	R/TA
1974	79 881	141 461	0.56	24 251	52 994	0.46
1975	121 101	140 492	0.86	25 532	60 650	0.42
1976	158 724	229 776	0.69	23 480	48 882	0.48
1977	177 955	251 112	0.71	21 135	45 646	0.46
1978	222 819	285 537	0.78	33 147	45 972	0.72

in agricultural research increased from 10.8% to 15.3%.[3]

If we compare the value estimated by Silva et al. (1980a,b) for agricultural research with the GNP for the same period, we find that these investments increased from 0.05% to 0.11%. If these values are compared with agricultural income, this ratio increases from 0.58% to 1.19% in the period 1974–78 (Table 7). These figures confirm that the government recognizes the need for investments in research and that the priority given to agriculture by the PND is shown by increases in expenditures on research.

This is confirmed by Table 2, which shows that expenditures in research are increasing faster than investments in technical assistance, although expenditures in technical assistance are higher. The ratio between agricultural research and technical assistance increased from 0.56% to 0.78% in 1974–78 at the federal level. Silva et al. (1980a,b) estimated that Brazilian expenditures on agricultural research and technical assistance in relation to agricultural production value were 0.70% for research and 0.94% for technical assistance. These authors estimated that, in 1974, the ratio between research expenditure and value of agricultural product in São Paulo was 0.81% and in Brazil 0.58%. According to Boyce and Evenson (1975) in countries with the same per capita income level as São Paulo this ratio was 1.83%; whereas in countries with per capita income similar to Brazil this ratio was 0.92%. Thus, expenditures in agricultural research have been less than what would be expected, especially when one considers the high returns that can be obtained from research.

Resource Allocation Criteria

Resource allocation criteria can be analyzed in two ways: by drawing information from people who make decisions upon the criteria; or by analyzing what has happened in the past, i.e., what criteria have been adopted in the past. However, although it is possible to describe the decision-making process, it is not possible to clearly establish the decision-making criteria.

Decision-Making Process

EMBRAPA's planning system until 1979 consisted of two stages. During the first, researchers received information on priorities. During the second, proposals were furnished by the researchers and submitted to the board of directors for final approval. The main feature of this system was that decisions at the administrative level prevailed over considerations coming from the research units.

The new system replaces this linear process with a circular model of agricultural research programing, where the decisions are taken "in loco," as a result of consensus between the participants, thus eliminating the power of veto of the central administration. In this new system, EMBRAPA's executives make decisions related to agricultural research by considering the policies of the Ministry of Agriculture, the PND, and PBDCT and the results obtained during the previous years. The board decides on the programs but the projects are decided upon by the researchers together with the technical assistants and farmers.

Priorities are established according to national interests. EMBRAPA is now concentrating its activities in the following areas: crops for exports; crops for food; crops for energy; problems related to more efficient use of inputs; and water management. In its planning, EMBRAPA considers the various regions not only in relation to national priorities but also in terms of the immediate needs of the farmers.

EMBRAPA has concentrated institutional, human, and material resources in the following fields: (1) research projects aimed at increasing natural agricultural productivity to meet domestic food demand, fulfill the needs of industrial development, and increase exports; (2) research to develop other

---

[3] The Third PBDCT has already been published but it does not give information about resources.

Table 3. EMBRAPA's investment in agricultural research by project, 1976–79 (figures in thousands of Cr$ at current prices). Source: DDM/EMBRAPA in Souza 1980.

	Investments			
	1976	1977	1978	1979[a]
Rice	15.157	24.224	34.196	49.124
Cotton	9.632	20.860	39.557	53.514
Beef cattle	30.417	57.309	71.406	92.972
Pork and poultry	5.293	6.030	8.319	17.217
Dairy	31.780	37.322	56.251	80.710
Beans	15.901	17.024	27.927	52.965
Fruits	24.458	32.975	46.110	76.424
Cassava	7.638	17.010	23.134	46.605
Corn	15.797	25.948	38.475	55.695
Vegetables	11.195	17.074	35.569	57.366
Soybeans	4.183	5.000	9.742	19.911
Wheat	3.348	2.207	3.023	5.831
Sorghum	7.123	13.051	20.228	22.434
Sheep raising	2.173	7.691	10.195	12.722
Soil	11.525	21.536	58.263	81.842
Genetic resources	4.293	6.541	6.559	9.190
Basic seeds	1.377	20.691	53.207	101.495
"Cerrados" program	24.510	30.111	46.452	61.736
Semi-arid tropics	11.626	20.318	27.887	53.681
Humid tropics	14.288	27.504	41.871	63.940
Food and agr. tech.	4.107	7.088	13.536	16.848
Rubber tree	12.613	18.027	26.448	24.674
Total	276.903	469.538	763.684	1202.952

[a] Budget for 1979.

Table 4. Personnel in the cooperative agricultural research system at the end of 1980. Source: DRH/EMBRAPA.

System components	Research	Research support	General administration	Total	
EMBRAPA	1553	3314	1902	6769	(48%)
State agencies	765	2459	1472	4706	(33%)
Integrated programs	617	1751	357	2725	(19%)
Total	2935	7524	3741	14200	

little-known and potentially important resources for the country, especially in the areas of the humid tropic, semi-arid areas, and "Cerrados"; (3) research to increase manpower and other inputs to improve income distribution; and (4) research aimed at reducing the Brazilian dependency on foreign fuel imports.

Analysis

Various attempts have been made to analyze the criteria for resource allocation to Brazilian agricultural research. One method of determining how this allocation has been done is to examine the number of publications and the subjects they cover.

The analysis done by Silva et al. (1979) showed that research on export products was more common than research on products for domestic consumption. If the number of publications on the principle export products (coffee, soybeans, sugarcane, cotton, citrus fruits, and cocoa) are compared with the number on the six most important food crops (rice, wheat, beans, potato, cassava, and corn) for the period 1927–77, it can be seen that the export products have received more than 60% of the research activity (Table 8).

Agricultural research is concentrated in the southeast region of Brazil (almost 80% of the total) and within this region, in the state of São Paulo.

Table 5. Number of research workers in EMBRAPA's graduate program 1974–79. Source: DRH/EMBRAPA.

	Place	Level	1974	1975	1976	1977	1978	1979	Total
From EMBRAPA	Brazil	M.S.	267	152	232	54	76	54	835
	Brazil	Ph.D.	20	7	3	4	2	16	52
	Abroad	M.S.	28	71	49	19	20	19	206
	Abroad	Ph.D.	19	28	21	15	33	30	146
Subtotal			334	258	305	92	131	119	1239
From other organizations (financed by EMBRAPA)	Brazil	M.S.	32	23	56	35	78	53	277
	Brazil	Ph.D.	3	1	—	—	2	1	7
	Abroad	M.S.	5	9	13	19	13	27	86
	Abroad	Ph.D.	8	8	14	28	9	8	75
Subtotal			48	41	83	82	102	89	445
Total			382	299	388	174	233	208	1684

Agricultural research in São Paulo represents 63% of the total for Brazil in the period 1927–77. It is only in the last decade, which corresponds to the creation of EMBRAPA, that agricultural research in Brazil (São Paulo excluded) has acquired real importance. Research in São Paulo traditionally has been channelled to export products, but in the last two decades food crop research has become more important. In the other regions that were studied, although it was observed that during the first decade research was oriented toward domestic products, significant changes took place and during 1970–77 export products received most of the agricultural research. In government institutions, research aimed at increasing land productivity prevailed.

Agricultural research in Brazil has been oriented by social and economic pressures. The guidelines established for Brazilian agricultural research seem to have been correct. The state of São Paulo, which had for a long time an economy tuned to foreign markets, concerned itself with research on export products; when the other states turned to a more commercial type of agriculture they also began to study these products. On the other hand, when a crisis in domestic supply occurred, the research system gave more emphasis to products that were important for the internal market.

Table 6. Personnel distribution within EMBRAPA. Source: DRH/EMBRAPA.

	Research	Research support	General admin.	Total
Cotton	36	92	34	162
Rice and beans	55	112	42	209
Sheep raising	22	56	25	103
Beef cattle	46	99	42	187
Dairy	60	196	57	313
Soils	68	49	43	160
Cassava and fruits	48	109	62	219
Corn and sorghum	51	207	82	340
Genetic resources	21	22	21	64
Rubber tree	27	63	26	116
Soybeans	49	116	34	199
Pork and poultry	32	52	21	105
Food and agricultural technology	28	33	30	91
Wheat	52	135	36	223
Total	595	1341	555	2491

Implications of Decision Criteria

This analysis of the Brazilian research system suggests that the method of resource allocation has been inconsistent. Recently EMBRAPA has tried to develop a system of decision-making; however, our research shows that this model is still not operating efficiently. Nevertheless, this does not mean that the research system is totally independent of what is happening in agriculture. On the contrary, there are many examples of quick response in the research system to the presence of problems in the agricultural sector (coffee rust, cotton "murcha," orange "tristeza").

The analysis suggests that the links between research and agriculture are established more clearly for a certain group of products, particularly export products and/or industrial raw materials. Products that are important components of the diet of the people and which are generally consumed raw or with minimal processing (rice, beans, cassava) have

Table 7. Public investment in agricultural research in relation to gross national product and agricultural income (figures in millions of Cr$).

	Agricultural research (AR)	GNP	AR/GNP (%)	Agricultural income (AI)[a]	AR/AI (%)
1974	382.2	713 336	0.05	65 657	0.58
1975	745.4	995 364	0.07	87 821	0.85
1976	1371.6	1 536 444	0.09	137 703	1.00
1977	2193.7	2 281 707	0.10	236 849	0.93
1978	3809.7	3 408 778	0.11	320 671	1.19

[a] Same value as agricultural product.

Table 8. Number of publications in agricultural research and agricultural research technology for six selected export crops and six selected food crops in Brazil, 1927–77. Source: Silva et al. 1979.

	1920–29	1930–39	1940–49	1950–59	1960–69	1970–77	Total
Export							
Coffee	23	38	71	132	163	457	884
Soybeans	1	5	4	12	25	189	236
Sugarcane	11	70	45	74	96	114	410
Cotton	12	50	38	48	115	63	326
Citrus	2	35	64	40	60	60	261
Cocoa	—	—	—	3	9	81	93
Subtotal	49	198	222	309	468	964	2210
Percentage	75.4	80.2	57.8	60.7	55.5	59.4	60.2
Food							
Beans	—	—	18	6	92	143	259
Rice	1	4	18	59	89	132	303
Wheat	7	1	14	15	17	207	261
Potato	5	15	51	51	80	30	232
Cassava	2	5	28	16	15	20	86
Corn	1	24	33	53	83	127	321
Subtotal	16	49	162	200	376	659	1462
Percentage	24.6	19.8	42.2	39.3	44.5	40.6	39.8
Total	65	247	384	509	844	1623	3672

received little attention. In other words, the expectation of future profits induces groups involved in industry and foreign trade to put pressure on the research system. They have been relatively successful in their efforts (see Pastore et al. 1976).

The important restriction in this system is food products that have market controls that make investments in modernization risky (Melo 1978) and over which the producers and consumers have little control. In other words, these products, for which the benefits of research could be potentially higher given their importance in the total consumption of the country, receive less attention in the research system because of government intervention and difficulties the producers face in trying to organize themselves.

In fact, the cooperative system has only been established for farmers working on export crops such as soybeans and food products such as poultry and vegetables, which are consumed by the higher income population.

The Brazilian experience suggests the difficulties encountered by a decision-making system when market mechanisms have limited use as signals to resource allocation due to the intervention of government agencies and an agricultural policy that tries to maintain low food prices. It also shows that the organization of producers is essential to make their voices heard and that exports and industrialization favour the establishment of these organizations. The major difficulty is how to make the voices of small producers and urban consumers heard.

Boyce, J.K. and Evenson, R.E. 1975. Agricultural research and extension systems. Department of Agricultural Economics, University of the Philippines at Los Baños, Laguna, Philippines.

Melo, F.B. Homem de. 1978. Agricultura Brasileira: incerteza e disponibilidade de tecnologia. São Paulo, Dep. de Economia e Adm., Univ. São Paulo. Tese de Livre Docência, mimeo.

Pastore, J., Dias, G.L.S., and Castro, M.C. 1976. Condicionante da produtividade de pesquisa agrícola no Brasil. Estudos Econômicos, 6(3), 147–181.

Silva, Gabriel S.P. da, Fonseca, M. Aparecida S. da, and Martin, Nelson B. 1979. Pesquisa e produção agrícola no Brasil. São Paulo. Agricultura em São Paulo, Ano XXVI, Tomo II, 175–253.

1980a. Os rumos da pesquisa agrícola e o problema da produção de alimentos: algumas evidências no caso de São Paulo. Revista de Economia Rural, 18(1), 37–59.

1980b. Investimento na geração e difusão de tecnologia agrícola no Brasil. Revista de Economia Rural, 18(2), 328–338.

Souza, I.S.F. de. 1980. Accumulation of capital and agricultural research technology: a Brazilian case study. Ph.D. Thesis, Ohio State University.

The Agricultural Research System in Malaysia: A Study of Resource Allocation

Mohd. Yusof Hashim[1]

Agricultural research is certainly not new to Malaysia, but its importance in the overall economic development of the country has only been fully appreciated in the last decade or so. During much of the colonial and postindependence period, research was largely concentrated on a single commodity, rubber. Because it was a major rubber producing country, it was logical for Malaysia to have given considerable emphasis to rubber research. Of the total cropped area of 3.85 million hectares, rubber occupies 54% of the area, oil palm 15%, rice 15.2%, coconut 8.3%, and other crops a total of 6.6%. Rubber contributes about a third of the gross national product and more than 50% of export earnings. The Rubber Research Institute of Malaysia (RRIM), established in 1925 to undertake rubber technology research, is one of the oldest and most widely recognized research organizations in the world.

Because of uncertain rubber prices on world markets and the realization that Malaysia needed to diversify her agricultural base to produce other agricultural products for local consumption and for export, research on other commodities was given more attention. In 1969, the Malaysian Agricultural Research and Development Institute (MARDI) was established and took over the research functions of agencies such as the Department of Agriculture, the Fisheries Department, the Veterinary Department, the Malaysian Pineapple Industries Board, the National Tobacco Board, and the Food Technology Division of the Ministry of Agriculture. MARDI thus became responsible for research on all crop commodities (except rubber), livestock, and freshwater fisheries. However, in September 1979, another agency, the Palm Oil Research Institute of Malaysia (PORIM) was established to undertake research on oil palm.

In addition to RRI, MARDI, and PORIM, there are a few other agencies that conduct research in agriculture and related fields: the Forest Research Institute (FRI); the Veterinary Research Institute (VRI, for research on animal health); the Fisheries Research Institute (FRI, for research on marine fish); and universities, in particular, University Pertanian, University Sains, and University Kebangsaan. Research is also conducted by the private sector, namely estates, to cater for its own research requirements.

Research for commodities other than rubber is crucial because Malaysia also produces rice, its staple food, vegetables, fruits, tobacco, and legumes mainly for domestic consumption and oil palm, cocoa, pineapple, and copra mainly for export. This realization of the importance of agriculture, not only as a source of foreign exchange but as a means of modernizing and improving the socioeconomic well-being of the rural masses, has made the government allocate higher proportions of its national development budget for agricultural development. In the Second Malaysia Plan (1971–75) about 26.5% of the budget was allocated to agricultural development. In the Third Malaysia Plan (1976–80) the allocation was 25.5%, whereas the percentage allocated in the Fourth Malaysia Plan is about 19%. The agricultural sector was expected to grow at the rate of 7.3% during the Third Malaysia Plan as compared with 5.6% during the period of the Second Malaysia Plan (Table 1).

In an effort to coordinate research and scientific activities within the country, a National Scientific and Development Council (NSDC) was established in 1976. The main function of NSDC is to promulgate basic policy guidelines on agricultural research and development. NSDC therefore has an influence on the directions of agricultural research programs and policies, but it has no authority or power to direct operational activities. Thus, NSDC has not been a major instrument to effect change in the

[1] Malaysian Agricultural Research and Development Institute, Bag Berkunci No. 202, Pejabat Post Universiti Pertanian, Serdang, Selangor.

Table 1. Growth of agricultural output in Malaysia, 1971–80 (1970 = 100). Source: Third Malaysian Plan.

	1971	1972	1973	1974	1975	1980 (Projected)	Average annual growth rate	
							1971–75	1976–80
Rubber	104.4	104.3	123.5	122.0	155.6	208.2	3.1	6.0
Palm oil and kernel	136.7	167.7	187.0	237.1	296.8	634.2	24.3	16.4
Saw logs	103.4	117.9	132.1	132.9	108.2	149.6	1.6	6.7
Rice	108.6	110.4	118.0	126.1	120.2	143.5	3.7	3.6
Coconut and copra	99.8	101.6	103.2	105.4	106.5	114.7	1.3	1.5
Pineapple	95.2	90.9	86.3	87.9	81.7	92.4	-4.0	2.5
Pepper	109.7	105.4	92.2	113.1	124.0	174.7	4.4	7.1
Tea	120.0	87.5	80.0	77.5	72.5	59.1	-6.2	-4.0
Fish	107.9	104.4	131.8	152.2	159.5	192.2	9.8	3.8
Livestock[a]	103.9	112.0	109.0	116.8	125.1	164.3	4.6	5.6
Miscellaneous[b]	104.7	115.2	118.4	123.1	132.5	190.2	5.8	7.5
Aggregate production index	106.8	112.6	126.2	132.5	131.5	186.5	5.6	7.3

[a] Includes beef from buffalo and oxen, mutton, pork, and poultry meat and eggs.

[b] Includes sago, tapioca, cocoa, coffee, sugarcane, groundnuts, maize, fresh fruits, tobacco, spices, food crops, and other minor crops.

research operations of the various research institutions, although all research institutions should fall under its umbrella.

The Rubber Research Institute (RRI)

The Rubber Research Institute of Malaysia (RRIM) is one of the three major units operating under the authority of the Malaysian Rubber Research and Development Board (MRRDB). The two other operating units are the Malaysian Rubber Producers Research Association (MRPRA) and the Malaysian Rubber Bureau (MRB).

The MRPRA performs research into the compounding, processing, properties, and uses of natural rubber and provides technical service for Britain and laboratory support for technical services in America and Europe. The MRB handles mainly technical advisory services and publicity and has offices in Australia, Austria, Germany, India, Italy, Japan, Netherlands, Spain, the USA, and the United Kingdom.

The MRRDB functions under the Ministry of Primary Industries, a ministry with responsibility for industrial export crops such as rubber, oil palm, pepper, and canned pineapple. For operational purposes, MRRDB is funded through a cess (research cess of 2.2¢/kg) collected from the sale of rubber by the government. This cess is placed directly under the control of MRRDB. The fund was about $35 million ringgit annually in the 1960s and increased to about $40 million ringgit annually in the 1970s. The MRRDB has exclusive authority in utilizing these funds for its activities in both research and development.

Although the MRRDB is responsible for the overall administrative policies and procedures, the RRIM is directly responsible for undertaking research into the production, processing, and manufacturing and marketing of rubber. The RRI obtains its operating funds directly from the MRRDB and thus avoids the problem of having to request funds from central agencies of the government. The institute has a technical senior staff of 233, a junior staff of 1160, and a budget of approximately $30 million. Table 2 shows the budget allocations for recurrent and capital expenditures for 1976–80.

The Malaysian Agricultural Research and Development Institute (MARDI)

Although MARDI was founded in 1969, it only became operational in 1971. The main function of the institute is to conduct scientific, technical, economic, and sociological research with respect to the production, utilization, and processing of all crops (except rubber and oil palm), livestock, and freshwater fisheries. In the early years, it concentrated its efforts mainly on the development of physical infrastructure, the training of staff, and the determination of future strategies. MARDI has a total of 26 research stations strategically located

Table 2. Recurrent and capital expenditures for 1976–80 (M$). Source: RRIM.

	Recurrent	Capital expenditure
1976	21 840 825	16 238
1977	23 118 580	1 358 396
1978	26 515 990	2 085 934
1979	28 034 485	2 126 437
1980	27 813 336	2 127 615

throughout Peninsular Malaysia and employs 440 research scientists and about 1600 junior support staff.

MARDI is a statutory body under the Ministry of Agriculture. In total, there are seven statutory bodies and seven departments under the Ministry of Agriculture. Organizationally, MARDI is structured along commodity lines (Fig. 1). The institute is governed by a board that comprises representatives of various central government agencies, the private sector, and political organizations. The governing board is responsible for administrative, finance, personnel, and policy matters. The institute submits its requirements for development and operating activities to the board and subsequently these are submitted to the central government agencies for final scrutiny and approval.

The scientific council is responsible in providing research guidance or directions to the institute. Composed of leading scientists from local universities and both public and private research organizations, the scientific council scrutinizes, monitors, and evaluates the research programs and activities of the institute. A number of advisory committees under the chairmanship of the members of the scientific council undertake detailed examinations of the research activities under the various commodities to ensure that the institute is sensitive to the needs of the country.

The institute is dependent on the government for financial support, both for recurrent and capital expenditures. It submits its financial requests to the Treasury and the Economic Planning Unit for its operating and capital estimates, respectively. Similarly, requests for personnel are submitted to the Public Service Department for approval before recruitment. The governing board then verifies or endorses the approved budget and personnel allocations for a particular year. Budget and personnel defence is normally done on a program by program basis and approval is given according to the programs. The allocations for both development and

recurrent expenditures for the last 5 years are indicated in Table 3.

The Palm Oil Research Institute of Malaysia (PORIM)

PORIM is a new organization established in September 1979. It took over research on palm oil that had been previously undertaken by MARDI. The primary objective of PORIM is to conduct and promote research on the production, extraction, processing, storage, transportation, marketing, consumption, and end uses of palm oil and oil palm products.

The institute is managed by a board that comprises representatives from the oil palm industry and government agencies that are appointed by the Minister of Primary Industries. The institute is financed by a research cess of $4/tonne of palm oil produced and is managed directly by the board, which is similar to the Governing Board of the Rubber Research Institute. Initially, the government provided a launching grant of $4.4 million to help the institute develop its infrastructure, i.e., research stations and laboratories. The private sector also provided support to help the institute become established. The institute currently has a total staff of 140 research scientists and technical subordinates. The estimated budget for the current year is about $12 million ringgit for both operating and development expenditures.

Universities

Universities, like other government statutory bodies, are semiautonomous institutions of higher learning under the Ministry of Education. They are governed by their respective University Councils in respect of all financial, personnel, and policy matters. However, as in the case of government statutory bodies, their requests for financial and personnel requirements must be submitted to the central government agencies for final approval because all universities are fully funded by the government.

Table 3. Recurrent and capital expenditures for 1976–1980 (M$). Source: MARDI.

	Operating	Development
1976	21 000 000	12 425 800
1977	29 425 500	15 579 000
1978	34 877 700	18 442 020
1979	40 271 000	16 313 050
1980	44 400 000	22 311 360

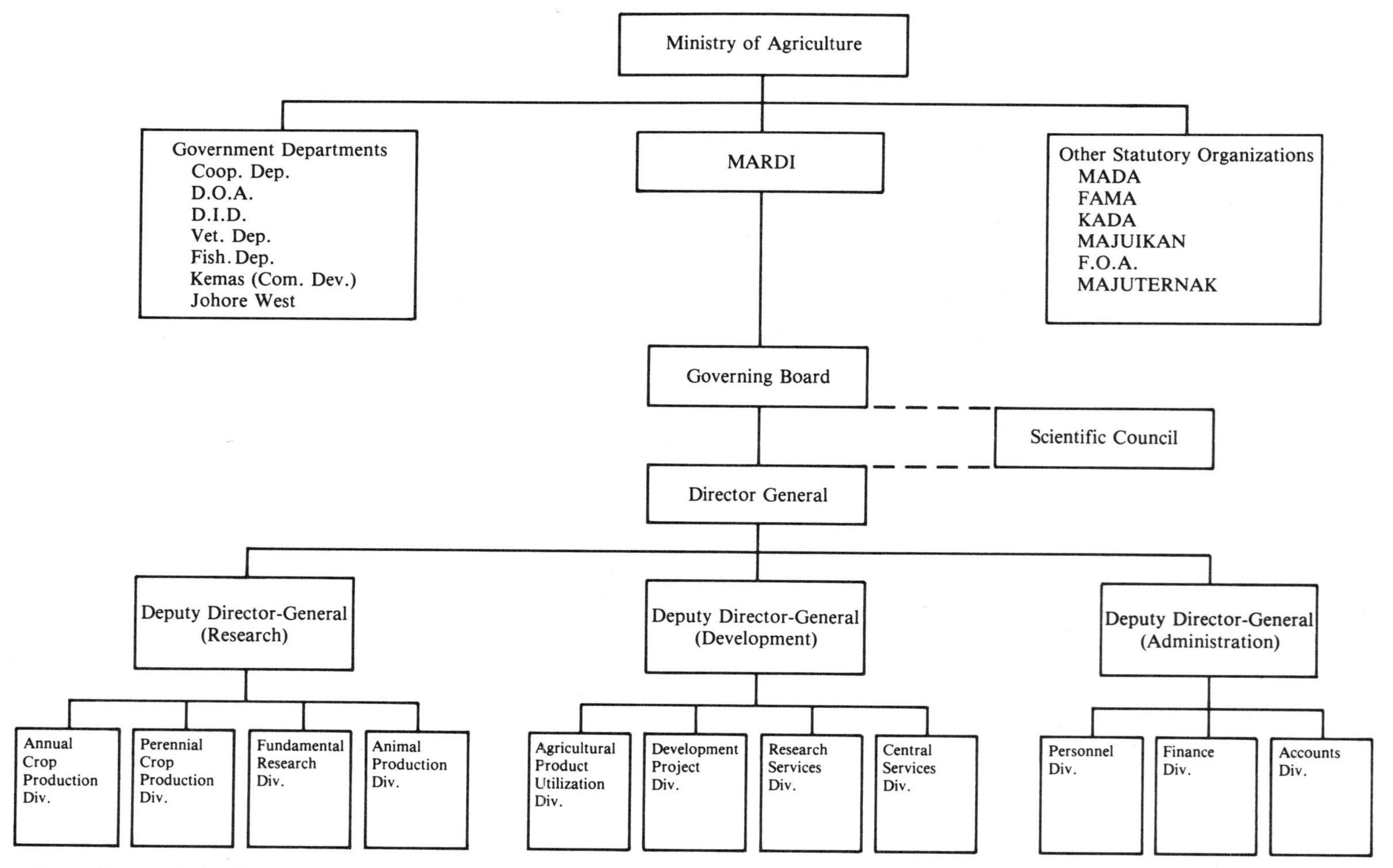

Fig. 1. The organizational structure of MARDI.

Other Government Agencies

Other government agencies/departments that conduct agricultural research include the Forest Research Institute (FRI), the Veterinary Research Institute (VRI), and the Fisheries Research Institute (FRI). These agencies operate like any other government department and are fully supported by the government in terms of financial and personnel requirements. Thus, they do not have any boards or councils to scrutinize their functions or activities. All requests for resources are submitted directly to the government's central agencies through their respective ministries.

Systems of Resource Allocation

From the discussion of the various research organizations, it is apparent that there are at least three different systems or models of operation prevailing in Malaysia. These can be described as:

(1) Research organizations that are autonomous and independent with regard to sources of funding. RRI and PORIM are in this group. They do not depend on the government directly for financial support. They obtain their funds from cesses levied by the government and these funds are directly controlled by their respective boards.

(2) Research organizations that are autonomous, controlled by their respective boards or councils, but must obtain financial support from the government treasury. MARDI and the universities are in this group. The boards or councils do not have the power to determine the amount of funds that can be made available to the organizations under them.

(3) Research organizations that depend directly on the government for their financial support without having to be scrutinized by other intermediary bodies such as a board or council. The FRI, VRI, and Fisheries Institutes are in this category.

Although Malaysia has a National Research and Development Council, it is only advisory in function and does not in any way have a direct influence on the operation of the research organizations, particularly in their efforts to secure financial and basic support. Therefore, there is no central body that coordinates or determines the resources to be allocated to research either on a commodity basis or as a composite of commodites for the country. The treasury, the central government agency, has the final authority to determine the amount of financial support to be given to research organizations (although not in the case of RRI and PORIM) based on annual requests. The allocation that is given is based on the merits of the requests made by each organization, not on a conscious effort to look at research needs as a whole.

The allocation of resources is an annual exercise in which estimates for financial support are prepared according to various codes of expenditure. The codes used in the estimates are: 1100 salary; 1200 allowance; 1300 overtime; 2100 travel and transport; 2200 transportation of materials; 2300 communication; 2400 utilities; 2500 rentals; 2600 printing; 2700 supplies and materials; 2800 maintenance and repairs; 2900 professional services; 3100 land; 3200 facilities; 3300 inventory; and 4100 training.

As an example, in the case of MARDI, the submission is made on a program by program basis. Table 4 indicates the categories and the amount asked for and finally allocated for 1981 after presentation to, and scrutiny by, the treasury. Normally, the heads of each division are asked to present their rationale/defence for the estimates based on the activiites envisaged for the year. Under normal circumstances, the increase in the budget estimates approved for each year should not increase more that 13% over the previous year's estimates.

A New Approach?

Is there a better approach or procedure that can improve resource allocation to agricultural research institutions in the country? Should there be a centralization of the decision-making process in resource allocation to all research organizations?

Because of current efforts of the government to develop the agricultural sector, particularly its emphasis on modernizing and increasing the socioeconomic status of the smallholder sector, it must be agreed that there is a need to increase allocations to research in the future. The present allocation

Table 4. MARDI's budget allocation for 1981.

Programs	Amount requested	Amount approved
Administration HQ	10 211 000	9 364 800
Station administration	20 739 900	14 245 600
Annual crops	5 697 200	5 972 400
Perennial crops	4 016 300	3 023 800
Livestock	4 147 900	6 112 300
Agricultural product utilization	4 351 300	4 649 500
Project development	2 920 000	2 507 200
Basic research	5 346 500	6 720 400
Research services	1 458 600	2 413 200
Central services	2 523 400	2 013 800
Subtotal	61 412 400	57 023 000
Increase in new salary structure	7 500 000	—
Total	68 912 400	57 023 000

to agricultural research is only about 0.2% of the GNP and this is far below the allocation made in many developed countries, which is normally between 2% and 4%. There is a need to build the research capabilities of the various research organizations in consonance with the need to develop the agricultural sector.

Due to the scarcity of resources, it would appear that their utilization would be more effective and efficient if they were centrally controlled, managed, and distributed. Such a measure would provide an opportunity for the government to study the research needs of the country in their proper perspective and to give priority to the areas that need immediate attention.

Centralization would also allow the government to monitor research inputs and outputs as a whole and enable it to distribute resources according to proper and rational demands. It would allow the government to build the research capabilities of the various institutes accordingly and in a systematic way rather than allowing the institutions to develop on their own, which can lead to disparity in the growth of different institutions. Centralization would also allow for coordination and sharing of resources, particularly expensive equipment and laboratory facilities, which would reduce the tendency for duplication and unhealthy competition. Centralization would also give the opportunity for the government to monitor and evaluate research in its contribution to the overall growth of the economy.

Concluding Remarks

The subject of resource allocation to agricultural research has become an important area of study in many countries. Malaysia is by no means an exception. In recent years, the subject has gained considerable attention because of the importance attached to the development of the agricultural sector.

On closer examination of the existing operations of the various research organizations, it can be discerned that there are at least three operational systems by which resources are allocated to research. In one system the organizations are rather independent in getting resources because the government has ensured that a certain amount of resources will be available by collecting a research cess for certain commodities. Another system allows autonomy in terms of policy and administrative actions but resources have to be sought from the government agencies. The other system does not allow any independence from the government in terms of administrative policy, and resource allocation and utilization.

There are, of course, some merits in allowing these systems to operate as they are. Organizations can make progress and develop themselves according to their own capabilities without having to worry about whether other organizations are well developed or whether research in other sectors of agriculture is given appropriate attention. In view of the scarcity of resources and the need to maximize productivity from the agricultural sector, it can be argued that if allocation were centrally coordinated, managed, and distributed it would be most efficient and effective. As well, the government would be in a better position to monitor and evaluate their impact if resources were allocated and utilized on a national basis rather than on the segmented basis presently practiced.

Human Resources in Agricultural Research: Three Cases in Latin America

Jorge Ardila,[1] Eduardo Trigo,[2] and Martín Piñeiro[2]

In recent decades, most of the countries of Latin America have made major efforts to train more and better researchers in the field of agriculture. Many of these efforts have received technical and financial support from a wide range of regional and international institutions as well as bilateral assistance from the developed countries.

Nevertheless, mounting evidence shows that a high proportion of the researchers trained in these programs have withdrawn from research activities in their native countries or, while not emigrating from the country, have abandoned their research programs and moved into different fields of activity. In some cases, withdrawal from research activities has taken place inside the organizations themselves as researchers are promoted to administrative and/or executive positions. This situation is the cause of growing concern in the countries themselves and in the international scientific community in view of its obvious negative effects on the development of adequate infrastructure for research in the countries of the region.

This paper will present the findings of an analysis of the problem at the Colombian Agricultural Institute (ICA), the National Institute for Agricultural Technology (INTA) in Argentina, and the National Agrarian University of La Molina (UNA) in Peru. These institutions were chosen because of their long-standing traditions in the field of agricultural research and because, in the last 20 years, they have made major efforts in the development of human resources.

The analysis is based on the hypothesis that excessive turnover of scientific personnel with high levels of training reflects complex, profound institutional problems related primarily to the bonds between the research activities and organizations, and the broader context of the societies in which they are operating.

This framework will be used in the analysis of the evolution of institutional models for the generation and transfer of technology, to which the organizations under study subscribe. The study will also examine training programs, the behaviour of highly trained research personnel, and the phenomenon and causes of migration.

Methodological Considerations

This conceptual approach proposes not only to describe and evaluate the process of training and migration of human resources, but also to describe these events within the context of the development and operation of National Research Centres.

Accordingly, three major working areas were identified. They cover the major types of problems related to professional migration that must be confronted by any organization (see Fig. 1). These areas can be identified as "conflict" areas that give rise to the phenomenon of migration. Each can be tied to concrete institutional factors and can be identified a priori with various types of migratory processes. Each working or "conflict" area, and the type of migration that characterizes it, is described below.

The first area involves the relationship between the institutions and the environment in which they operate. This includes "upward" ties with the rest of the public apparatus and other social organizations, as well as "downward" relationships with the target community (clientele). The institutions must design a strategy to maximize areas of contact in both directions to reach a suitable level of recognition and feedback for reaching their objectives efficiently. These "conflicts" have been seen to produce migratory phenomena of a general or indiscriminate nature. In other words, the processes generally affect professionals with no distinction for specialization or placement in the organization.

---

[1] Research Specialist, PROTAAL Project, IICA Office in Colombia.

[2] Research Specialist and Coordinator, respectively, of CIGTAT, IICA Headquarters, San Jose, Costa Rica.

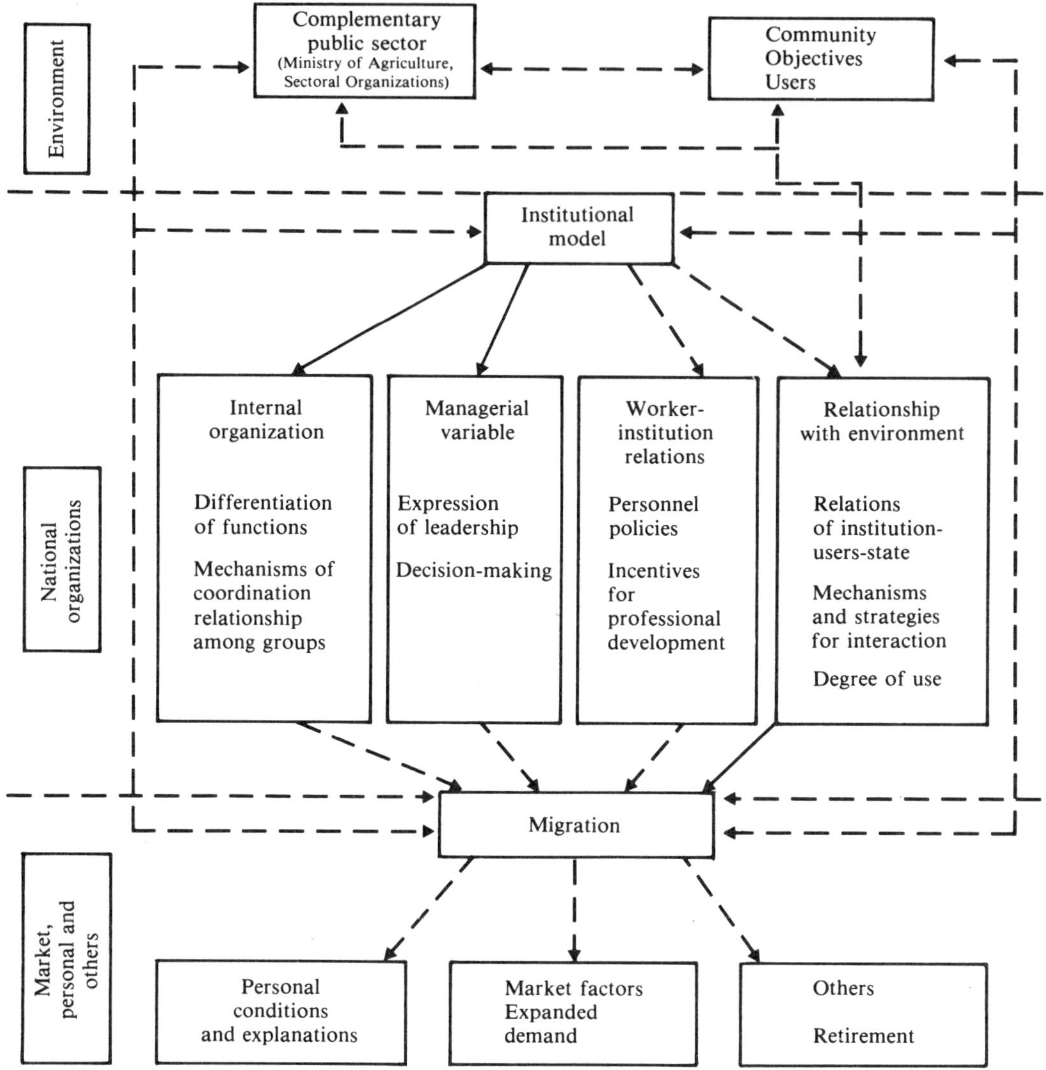

Fig. 1. General outline of variables related to the migration of qualified technical personnel.

These conflicts tend to coincide with relatively low levels of recognition for research activities, and therefore lead researchers to emigrate from their home country.

The second "conflict" area is internal to the institution and involves the form, nature, and evolution of the working relationships among the various components of the institution. Every organization at any given moment has its own division of work on which it bases its administrative operations and the execution of its strategies for reaching its objectives. This structure is not permanent over time. Rather, it is normally subject to frequent change, stimulated either by a loss of ability to reach institutional goals, or as a result of shifting power relations among the various sections of the organization. This situation brings about changes in institutional policies and in the status of technical personnel inside the institution and can be linked to certain migratory phenomena, particularly of the selective type, which affect only certain elements of the technical staff of the organization. These processes can give rise to a particular phenomenon that can be called "internal bureaucratic migration," wherein changes occur on certain levels, pushing specialists further and further away from their areas of specialization. The decisions for such changes are dominated by questions of institutional policy or administrative efficiency, which take precedence over technical considerations.

Finally, a third "conflict" area involves the relationship between the specialist and the institution. These relationships are determined by personnel policies and their effect on the development of technical skills and the work environment. Conflicts in this realm generate migratory processes with the characteristics described above, varying with the intensity and magnitude of the conflicts. The migratory process becomes more generalized and indiscriminate if there are problems concerning salaries or professional development. Such problems include the lack of training programs and the inability to attend technical meetings.

In view of these general concepts, the project included two phases of analysis. The first sought secondary information, from the research organizations themselves, for an analysis of the nature and development of institutional models for the generation and transfer of technology and for training programs. It also sought to describe and quantify the migratory phenomenon. The second phase was a survey of former employees of the institutions to identify and analyze the causes of their migration. The survey was designed to gather information on the variables associated with each of the three areas of conflict.[3]

The Development of Institutional Models for the Generation and Transfer of Technology

The Institutional Model

The gradual changes in the institutional models of the three cases under study clearly took different directions based mainly on changes in the location of research functions as an element of the organizational–functional scheme adopted.

In the case of the ICA, three clearly defined stages of institutional arrangements can be distinguished, each with its own significance for the location of research functions in the organization. During its first period, from 1962 to 1967, ICA's fundamental activity was agricultural research, backed up by extension and education on various levels. The institute's work was to support agrarian reform programs, with the objectives of improving the income of poor campesino farmers, increasing the stability of small farming units, and expanding the food supply by improving yields.

In 1968, overall agrarian policies, which had initially sought to change the structure of land tenure, were reoriented to promote technological change more intensively, and thus increase the efficiency of commercial agriculture and provide greater technical assistance for small-scale campesino farmers. At ICA, these changes took the form of new functions to control production factors and promote production, with special emphasis on livestock. The administration was expanded to the national level and the central organization was reproduced in eight regional centres. These modifications altered the original research and extension model, which gave way to an organization for promoting production, of which the generation of technology is only one part.

In 1973, as a result of a new effort to overhaul state agricultural policies, ICA underwent a reorganization that affected its organizational structure, its objectives, and its target population. Technological change continued to be viewed as an important strategic tool, but now it was viewed as only one part of the work to support core groups of small-scale farmers. This work included everything from credit to associative forms of production. ICA maintained its earlier functions and was given responsibility for conducting specific rural development projects within national programs for Integrated Rural Development, in which it handled the technological elements involved. Thus, after establishing itself as an institution giving high priority to research and the complementary fields of education and extension, by the end of the seventies it had become a multipurpose organization, essentially a state agency in charge of agricultural development with the support of research.

INTA in Argentina has undergone a pattern of development different from that of ICA. The major difference is that the original system, whereby research was integrated into, and complemented by, extension and educational efforts, has never been significantly altered. This model has been protected by provisions of the original legal system that created INTA, which prohibited the assumption of any additional unspecified functions.

In Peru, the National Agrarian University (UNA) shares agricultural research activities with other state agencies (such as the recently created INIA,

---

[3] For greater detail on the methods used in the study, see Sistemas nacionales de investigación agropecuaria en América Latina: análisis comparativo de los recursos humanos en países seleccionados. Ardila, J., Trigo, E., and Piñeiro, M. PROTAAL. Document No. 46, IICA, Office in Colombia, Bogota, March, 1980.

formerly known as the DGI).[4] Its operational model can basically be defined as that of a university centre dedicated to agriculture, whose teaching work is supported by research. In addition, through its postgraduate program, it is recognized as the country's most important centre for training researchers.

In this framework, research activities play a major role, but they are subordinate to academic endeavors and are structurally responsive to certain priorities of the university. Research funding is largely dependent on the outside initiative of public and private organizations, with which it holds contracts for conducting specific projects. This is its principal operational system.

The general fields of competence of the UNA have undergone no substantial changes since the university was created. Nevertheless, official agrarian policy in recent years has prompted shifts in research priorities to benefit community enterprises as a major recipient, downplay private enterprises, and reorient work toward basic consumer goods for the country.

The Allocation of Resources for Agricultural Research

An examination of the budgetary trends for the three organizations in terms of 1977 dollars revealed that ICA and INTA underwent similarly high levels of budgetary growth until 1970. At that point, they took separate directions. ICA began to show a negative trend, while INTA experienced sharp swings from one year to the next. By contrast, the UNA showed a certain stability in budgetary resources.

In the first half of the seventies, budgetary variations were greater; however, the highest budgets of the entire period under study were available during this period for ICA (1972) and INTA (1974). The UNA experienced general stability, maintaining similar levels over time, although well below the allocations received by INTA and ICA. Finally, after 1975, the budgets of ICA and the UNA dropped, while after 1976, INTA went into a period of recovery.

In the particular case of ICA, the 1968 reforms brought significant budget increases. However, the 1973 reforms led to general cuts that ushered in today's problems of heavy budget deficits.

Of particular note in INTA is a crisis in 1975–76, characterized by sharp cutbacks in resources. For the UNA, the 1974 reforms brought initial improvements that have been gradually falling off since 1976. These budgetary shifts cannot be given a direct interpretation, especially if budget allocations are viewed as indicators of the type and degree of support received by the institutions. In this sense, it is important to note the origin of the resources. The ICA and UNA funds come from the National Budget and therefore are a direct measure of the support being received by the organizations. In the case of INTA, on the other hand, funds are received as a percentage of agricultural exports and varying degrees of available funds do not reflect the degree of support as much as the current state of exports.

Table 1 gives figures on fund availability per specialist, as a measure of the real operating capacity of the organization. Due to the diverse nature of data sources, the information is not strictly comparable from one country to another. In the case of INTA, the total professional budget provided the figures. For ICA, figures are given for specialists with postgraduate degrees; the UNA figures give the budget per faculty member. However, the table does give some idea of the situation for each organization.

Available resources for INTA, after an initial fall due to the sharp increase in personnel, show a high level of stability, varying only in the 1975–77 period, which appears to be clearly atypical. By contrast, sharp swings occurred in the ICA. Until 1972, the budget per specialist grew steadily, but after that year it began to slump until, by 1978, it had fallen to only one-fourth of the 1972 level.

The UNA has also experienced sharp variations in funds per faculty member, which reached a high in 1967 and then dropped off almost without interruption, except for small rallies in 1975 and 1976.

Development of Training Programs

During the sixties, when agricultural research organizations began to take shape in the countries of Latin America, postgraduate training progams were given top priority. Because there were no postgraduate programs on the national level, these early efforts looked to foreign universities, with strong support from international organizations and foreign governments. In time, this approach was complemented and partially replaced by local efforts to create national postgraduate programs in the agricultural sciences.

These national programs (at the Master's level) sought to upgrade training in certain areas, attuning it to the local conditions of the countries in the region. In addition, they were set up as mechanisms for institutionalizing training efforts funded with national resources, once the foreign aid programs

---

[4] This is an important distinguishing factor because in Argentina and Colombia ICA and INTA have a formal monopoly over research activities in the public sector.

Table 1. Number of professionals and average budget per professional (in thousands of 1977 U.S. dollars) for 1966–78 in ICA (Colombia), INTA (Argentina), and UNA (Peru).[a]

	ICA		INTA		UNA	
	Number	Budget	Number	Budget	Number	Budget
1966	60	0.083	70	0.280	68	0.052
1967	90	0.141	87	0.288	70	0.078
1968	100	0.138	105	0.258	77	0.092
1969	107	0.195	120	0.215	90	0.055
1970	123	0.224	133	0.224	104	0.027
1971	142	0.197	141	0.175	123	0.039
1972	163	0.225	169	0.141	124	0.038
1973	192	0.135	193	0.169	126	0.038
1974	256	0.110	209	0.187	128	0.033
1975	311	0.096	221	0.138	127	0.026
1976	336	0.088	197	0.131	120	0.045
1977	371	0.075	195	0.149	117	0.050
1978	383	0.062	189	0.181	112	0.037

[a] Source: Developed from information provided by the institutions.

had been phased out. Thus, the ICA Graduate Programs emerged in Colombia (PEG) and the Graduate School of Agricultural Sciences appeared in Argentina (INTA-Castelar). The Graduate Program of La Molina was created as the only case in which training was to be provided by the university itself, in contrast to INTA and ICA, where from the very beginning, training programs were located in the research organizations.

The overall growth and development of the training programs in each organization varied. In ICA, the training program was divided into three components. To begin with, training abroad generally received a high percentage of foreign assistance. The second part was the ICA-ICETEX (Colombian Institute of Technical Specialization Abroad) agreement to replace training programs funded by international organizations, and under which ICA contributed funds to ICETEX for administration. Finally, the PEG (Graduate Study Program) was oriented toward creating national capabilities and was complemented with the ICA-ICETEX agreement in those areas in which national capacity was inadequate. Under the first area, which consisted essentially of scholarships from foreign organizations and governments, ICA trained 302 specialists, the majority of whom studied in the United States and in Chapingo and Monterrey, Mexico. The most significant contributions to this program were received from the Rockefeller Foundation (130 scholarships) and the University of Nebraska (139 scholarships). Additional participants included the Ford Foundation, the Kellogg Foundation, and USAID. The ICA-ICETEX agreement went into effect in 1971, at which time the international cooperation had nearly

ended. Under this agreement, priority was given to Ph.D. training. It did not have the expected results, however. In the first place, it went into effect during the years when ICA's budgetary and financial crisis was beginning; in the second place, most of the scholarships were used by administrative personnel instead of technical specialists, as the growth of the organizations was felt to require more efficient administration. Fifty-four members of the professional staff were trained through this channel.

In 1967, the PEG was created by agreement between the ICA and the National University, making use of the specialists the ICA had trained abroad, and this guaranteed a steady source of faculty members. By 1971, the program had become the major centre of specialized studies. By 1978, 274 technical specialists had been trained at the Master's level, and most of them were working for ICA. In recent years, the program has declined sharply, as have the other two training components (international agencies and the ICA-ICETEX agreement). Thus, at the present time there is no training structure that responds to the needs of the organization.

At INTA, a total of 314 technical specialists were trained. Of these, 87 were funded by international organizations, beginning with the Ford Foundation, which was later joined by USAID, the Rockefeller Foundation, the British Council, the French Agricultural Research Institute, FAO, and others on a smaller scale. The second component on this level was the national postgraduate program, which began to operate with the 1964 founding of the Graduate School for Agricultural Sciences. This school received fresh momentum in 1967 with the signing of an agreement that expired in 1975. The agreement was not renewed, and at present, only small-scale activities are taking place in certain fields (animal production and plant improvement).

By 1978–79, INTA, like ICA, had no effective alternatives for training technical personnel. This was due to the state of the national graduate program and to the fact that most of the support from foreign institutions had been terminated during the early seventies.

The La Molina Agrarian University is an academic centre of renowned excellence. It had begun providing its specialists with technical training, prior to the creation of the graduate program, with support from international institutions. A total of 190 specialists graduated in this way, with major assistance coming from USAID, the Rockefeller Foundation, the British Council, and, to a lesser extent, 10 other institutions.

The UNA graduate program, in addition to training faculty members, addresses the national demand for specialists and researchers. This program has trained 38 specialists, or 15% of the total personnel trained at the postgraduate level. Currently, as in the cases of INTA and ICA, this program is experiencing severe difficulties, and its importance is falling off rapidly. Nevertheless, funding possibilities still exist through support that has recently become available from the IDB. Table 2 shows the trajectory of the training programs, in terms of participants and budgets, taking into consideration the use of both national and foreign resources. It also shows figures for attendance in studies abroad compared with studies in the countries themselves.

Until 1973, the three institutions showed a growth in programs (Table 2). After that year, only ICA continued to grow, thanks to the effects of the PEG, a trend which lasted through 1976. These data bring to light the magnitude of the ICA program, with a total of 630 specialized personnel, compared with 314 at INTA and 260 at La Molina.

As expected, it was found that, with the establishment of graduate programs in these institutions, the percentage participation in training programs abroad fell rapidly. This was due not only to the expansion of home-based alternatives, but also to reductions in absolute levels of foreign funding.

Finally, Table 2 shows the costs of training programs, including expenses for travel, tuition, and housing for the scholarship recipients in their chosen universities, as well as salary compensation during the study period. It also shows clearly the magnitude of the ICA program, with a cost of nearly 20 million dollars up to 1977. The other two institutions show much lower sums: 5.2 million for INTA and 2.9 million for La Molina. In all three cases, foreign funding of these programs was an important element. In La Molina it totalled 94.5%, whereas in ICA it was 50% and in INTA 37%. It should be noted that the figures for La Molina do not include salary outlays during the period of study because most of the specialists studying abroad under scholarships were on leave without pay and thus gave up all salary benefits during that period.

To determine the priority given to professional improvement, training expenses can be figured into the total budget of the institutions. Thus, in 1965 INTA spent 1.11% of its funds on training, whereas in 1978, only 0.001% of its total budget went into this item. In 1966, the ICA spent 4.5% of its budget on training, and by 1978, it spent only 0.91%. No information from the UNA was available for calculating this factor. These rates show that training has never been a high priority item and, that with the passage of time, it has lost even the little importance it once enjoyed. This is confirmed by the fact that, when outside funding terminated with the expiration of the respective agreements, it was never replaced, and training alternatives gradually disappeared.

Overall Human Resource Patterns

Data are given below on the changes over time of personnel with postgraduate degrees. This information is included in order to trace the trajectory of the staff as a result of the combined effect of migration and personnel training policies. In addition to overall information, the data have been broken down by M.S. and Ph.D. degrees.

Overall Changes in the Institutions

The number of staff with postgraduate degrees in each of the institutions was very similar during the sixties, and all three institutes showed a steady increase in numbers. However, differences began to emerge in the early seventies. In the first place, the UNA staff began to stagnate, and by mid-1975, a slow, steady annual decline had begun. The total staff of INTA continued to grow during the first half of the decade, but in 1976, with the crisis experienced in the institution, it began to fall off. Finally, the ICA's staff grew throughout the period, as a result of the scope of its training programs.

In spite of these differences from one institution to the next, certain common traits can be pinpointed. This is particularly true in terms of the steady growth of staff prior to the seventies, and the appearance of a downward trend during the second half of the decade.

Table 2. For ICA, INTA, and UNA, number of new students per year receiving postgraduate training (percentage of total who received training abroad in parentheses) and total cost of training program (in thousands of 1977 U.S. dollars) (percentage of cost covered by foreign funds in parentheses).[a]

	INTA		ICA		UNA	
	New students	Cost	New students	Cost	New students	Cost
1960[b]	7 (100)	134 (67.3)[c]	5 (100)	260 (36.7)	33 (97)	496 (91.3)
1961	17 (100)	182 (54.4)	9 (100)	365 (45.8)	7 (100)	96 —
1962	9 (100)	244 (54.1)	17 (100)	583 (52.3)	11 (100)	153 (91.3)
1963	18 (100)	235 (62.2)	14 (100)	666 (43.2)	17 (88.3)	232 (85.6)
1964	23 (73.9)	323 (57.5)	10 (100)	670 (48.0)	19 (94.7)	227 (94.5)
1965	15 (86.7)	371 (45.6)	11 (100)	501 (53.5)	15 (86.7)	148 (87.5)
1966	22 (86.4)	193 (38.3)	22 (100)	522 (55.8)	13 (84.6)	134 (92.9)
1967	34 (100)	406 (51.5)	24 (66.7)	808 (49.2)	27 (96.3)	344 (99.4)
1968	28 (64.3)	570 (51.5)	35 (80.0)	933 (49.6)	24 (75.0)	248 (97.9)
1969	23 (91.3)	437 (52.5)	40 (80.0)	1288 (52.1)	16 (87.5)	187 (95.9)
1970	21 (76.2)	526 (45.2)	51 (70.6)	1725 (50)	20 (80.0)	213 (95.9)
1971	39 (46.2)	553 (42.3)	37 (78.4)	1530 (47.7)	10 (70.0)	91 (93.5)
1972	24 (66.7)	346 (42.1)	110 (46.1)	1830 (39.9)	10 (90.0)	105 (98.1)
1973	24 (33.3)	258 (45.5)	96 (45.8)	2356 (31.2)	11 (54.6)	74 (91.8)
1974	4 (100)	263 (32.4)	57 (29.8)	2083 (23.6)	13 (61.5)	84 (83.9)
1975	1 (100)	120 (38.9)	53 (3.8)	1441 (17.7)	7 (85.7)	75 (99.1)
1976	2 (100)	33 (49.9)	28 (17.9)	991 (14.7)	6 (66.7)	42 (97.5)
1977	1 (100)	8 (39.8)	7 (100)	659 (12.7)	1 (100)	— —
1978	— —	1 (58.0)	4 (25.)	411 (23.0)	— —	— —
Total	314 (76.7)	5203 (49.0)	630 (56.5)	19621 (37.7)	260 (85.4)	2949 (94.5)

[a] Source: Developed on the basis of information provided by the institutions.
[b] This figure includes students from previous years.
[c] Includes students financed with national funds but studying outside the country.

Patterns of Change by Level of Training (M.S. and Ph.D.)

In all the institutions, figures on staff members trained at the M.S. level followed patterns very similar to those of the overall staff. This is simply because the majority of the postgraduate personnel are in this category. Thus, the UNA underwent a period of stagnation between 1971 and 1976, and a drop-off after 1976. At INTA, the levels of growth held steady until 1975 and then began to slack off. Finally, growth in ICA continued throughout the period under study, peaking in 1974 and 1975 (as a result of the PEG).

At the Ph.D. level, the pattern is completely different. Until 1973, ICA showed strong growth that drew to a standstill after that year. This pattern, which contrasts sharply with the experience at the M.S. level, produced a striking change in the overall staff composition, with a sharp increase in the M.S./Ph.D. ratio. INTA's Ph.D. staff grew considerably from 1968 through 1975, dropping off in 1976. Until 1973, the UNA experienced strong growth. However, after 1974, the bulk of the resignations occurred at the Ph.D. level, so that by 1978, the staff had fallen to almost half the 1973 level.

Migration: Quantification and Analysis

Resignation Trends in Absolute and Relative Terms

Table 3 gives the findings of a study of the volume and historical development of the migration of personnel with postgraduate training. In general, on the basis of these findings, it can be stated that several distinct stages have taken place. The first lasted until 1975 and was characterized by low levels of migration. During the second stage, personnel turnover was relatively normal (around 6.0%). This stage took different forms in the three institutions. At INTA, it lasted from 1965 through 1978, with the exception of 1976, when migration rose abruptly. ICA, by contrast, experienced low levels of turnover from 1965 to 1970, at which time it began rising, only to fall again toward the end of the period, in

Table 3. For ICA, INTA, and UNA, number of resignations (in absolute values) of those with postgraduate training. The annual turnover rates[a] for postgraduate personnel are given in parentheses.[b]

	ICA			INTA			UNA		
	Total	M.S.	Ph.D.	Total	M.S.	Ph.D.	Total	M.S.	Ph.D.
1960	— (0)	— (0)	— (0)	— (0)	— (0)	+[c] (0)	— (0)	— (0)	— (0)
1961	— (0)	— (0)	— (0)	— (0)	— (0)	+ (0)	— (0)	— (0)	— (0)
1962	— (0)	— (0)	— (0)	— (0)	— (0)	+ (0)	— (0)	— (0)	— (0)
1963	— (0)	— (0)	— (0)	— (0)	— (0)	— (0)	— (0)	— (0)	— (0)
1964	— (0)	— (0)	— (0)	— (0)	— (0)	— (0)	— (0)	— (0)	— (0)
1965	1 (2.4)	— (0)	1 (6.9)	— (0)	— (0)	— (0)	— (0)	— (0)	— (0)
1966	1 (2.1)	— (0)	1 (5.7)	2 (3.2)	2 (3.3)	2 (0)	— (0)	— (0)	— (0)
1967	— (0)	— (0)	— (0)	1 (1.3)	1 (1.3)	— (0)	2 (2.9)	2 (3.5)	— (0)
1968	4 (6.2)	1 (2.5)	3 (12.8)	1 (1.0)	1 (1.1)	— (0)	2 (2.7)	1 (1.7)	1 (7.7)
1969	3 (4.1)	2 (4.1)	1 (4.0)	4 (3.5)	4 (3.8)	— (0)	1 (1.2)	1 (1.5)	— (0)
1970	3 (3.4)	3 (5.0)	— (0)	7 (5.5)	7 (6.1)	— (0)	4 (4.1)	3 (4.1)	1 (4.3)
1971	12 (10.4)	7 (8.7)	5 (11.6)	9 (6.6)	8 (6.7)	1 (5.7)	4 (3.5)	4 (4.8)	— (0)
1972	14 (9.6)	10 (9.4)	4 (10.3)	2 (1.3)	2 (1.5)	— (0)	12 (9.7)	11 (12.6)	1 (2.8)
1973	15 (8.5)	9 (6.8)	6 (13.6)	4 (2.2)	3 (2.0)	1 (3.2)	6 (4.8)	5 (5.8)	1 (2.6)
1974	28 (12.5)	20 (11.3)	8 (17.2)	3 (1.5)	1 (0.6)	2 (6.3)	14 (11.0)	9 (10.3)	5 (12.7)
1975	47 (16.5)	36 (15.3)	11 (22.9)	4 (1.9)	1 (0.6)	3 (9.0)	9 (7.1)	3 (3.3)	6 (16.7)
1976	30 (9.3)	22 (8.1)	8 (15.8)	30 (14.4)	20 (11.3)	10 (31.3)	9 (7.3)	6 (6.5)	3 (9.5)
1977	20 (5.7)	15 (5.0)	5 (9.7)	3 (1.5)	2 (1.2)	1 (3.4)	8 (6.8)	6 (6.7)	2 (7.0)
1978	16 (4.3)	15 (4.6)	1 (1.9)	7 (3.7)	6 (3.7)	1 (3.5)	14 (13.4)	10 (11.8)	4 (16.0)

[a] Number of personnel changes which have taken place in the corresponding period. Calculated on the basis of resignations over average annual staff.
[b] Source: Developed from information provided by the institutions.
[c] Plus sign indicates there were no Ph.D. personnel in these years.

1977 and 1978. Finally, UNA generally had low levels of turnover until 1971, after which migration rose to high levels.

In general, the seventies can be described as a period of high migration by postgraduate personnel. Peaks were reached in 1974–76 at ICA, in 1976 at INTA, and in 1978 at UNA.

A breakdown of this information by levels of specialization indicated a higher percentage of migration at the Ph.D. level in all three institutions. Worse yet, during the periods of highest migration, the difference between the number of researchers with a Ph.D. and those with a M.S. was accentuated, which indicates greater sensitivity at the Ph.D. level to the factors prompting the resignations.

Analysis of the Migration Balance

We have now briefly analyzed the trends of the total postgraduate staff, as well as data on patterns of resignation in absolute and relative terms. The annual difference between hirings and resignations (known as the migration balance) was also examined for 1960–78. This information showed that during the early period, 1960–69, all three institutions maintained consistently positive balances, and it was a time of net staff development.

From 1970 on, sharp differences emerged from one institution to another. From 1970 to 1974, ICA had a rapidly growing migration balance, in spite of a high level of resignations. This is because of the extensive training that took place during those years. The balance then fell off sharply until, in 1978, it approached negative levels that were repeated in succeeding years, as resignations outnumbered new hirings. At INTA, a positive balance was maintained from 1970 through 1975. However, beginning in 1976, the balance became negative, as occurred in the UNA. In spite of differences of magnitude, it can be stated that from 1974 on, the situation grew worse for all the institutions under study.[5]

This analysis reveals the interaction between the migration balance and training policies as determining elements in total staff levels of trained personnel. Thus, in spite of the apparent drop-off in the rate of resignations that occurred in ICA, the staff situation tends to grow worse because training programs were cut back.

---

[5] Information obtained after this analysis had been completed indicated that the migration balance for ICA from 1979 to 1980 was also negative at the Ph.D. level, and that the rate was higher than that of INTA and the UNA after 1974.

Analysis of Turnover Rates

An important consequence of the extensive migration and reduction of training programs in the seventies was the increasing incidence of young, inexperienced technical personnel. There can be no doubt that this has a negative impact on the effectiveness of research activities, in which long periods of time are often required to obtain results of quality and impact. Thus, interruptions or changes among the qualified personnel can have a negative effect on the outcome.

Therefore the personnel turnover rate was calculated for each institution, by year and by longer periods of time. This turnover rate gives a percentage of the number of times during a given period that personnel in the institution had to be replaced (Table 3).

In the beginning, the rates maintained normal levels until approximately 1970. After 1971, ICA and the UNA experienced considerably higher turnover rates.

However, the rates at INTA remained low except in 1970–71 and 1976. If figures are broken down by level of training, the three institutions generally show higher rates of turnover at the Ph.D. level than among the M.S. group, a phenomenon that has been increasing with time. ICA particularly stands out for its high levels of turnover, particularly at the Ph.D. level where, during the period under study, the entire staff was replaced almost twice and the M.S. staff was replaced 1.26 times.

This situation strongly influences the profitability of investments in training. High turnover rates mean that the specialists are spending very little time in the organization and, as a result, possibilities of recovering the initial investment are minimal. The Ph.D. specialists require the highest investment, and the higher turnover rates at this level only make the ratio worse.

High turnover rates imply a certain amount of institutional disorganization (continuous personnel changes, etc.) that undoubtedly has a strong effect on the productive capacity of the institutions.

Migration by Speciality and Area

In this analysis and quantification of migration, "speciality" refers to postgraduate training (genetics, entomology, pathology, etc.) and "area" covers groups of specialities that deal with similar problems (such as plant protection, including plant pathology and entomology). Tables 4 and 5 give the information from this analysis.

Table 4. For ICA, INTA, and UNA: specializations as a percentage of total postgraduate personnel and estimated rates of gross migration by specialization (1960–1978).[a]

Specialization	ICA		INTA		UNA	
	% of total	% emi-gration	% of total	% emi-gration	% of total	% emi-gration
Crop Production	*39.3*		*34.3*		*25.2*	
Plant improvement[b]	12.8	89.7	7.3	43.7	2.6	75.0
Crop production	8.6	13.6	2.5	0.0	—	—
Plant pathology	4.9	64.7	3.8	9.1	3.4	80.0
Plant physiology	3.8	120.0	2.9	200.0	4.5	33.3
Entomology	3.5	33.3	2.2	16.7	1.9	0.0
Pasturage	—	—	5.4	6.2	—	—
Soils	5.7	—	10.2	18.5	10.2	28.6
Irrigation	—	—	—	—	2.6	250.0
Livestock	*20.2*		*17.2*		*13.6*	
Animal genetics[c]	4.9	21.7	2.2	16.7	4.9	44.4
Nutrition	4.3	47.1	3.2	25.0	3.4	12.5
Animal husbandry	4.2	118.2	—	—	—	—
Veterinary medicine	3.8	29.4	—	—	0.4	0.0
Animal production	—	—	10.2	10.3	1.1	0.0
Animal pathology	3.0	30.8	1.6	0.0	—	—
Biochemistry	—	—	—	—	3.8	—
Extension and Development	*14.4*		*12.1*			
Extension and communication	8.2	56.7	12.1	52.0	—	—
Rural development	6.2	12.5	—	—	—	—
Other	*9.0*		*14.0*		*18.1*	
Agricultural economic	5.9	88.9	14.0	214.0	8.7	109.1
Economic (planning)	3.1	125.0	—	—	6.0	433.3
Business admin.	—	—	—	—	3.4	800.0
Total	82.9	51.4	77.6	43.4	56.9	84.1

[a] Source: Developed from information provided by the institutions.
[b] At INTA, includes plant genetics.
[c] At ICA, includes animal microbiology.

All three institutions showed a predominance of crop production-related specialists in comparison with the personnel involved in livestock activities. Because of its academic nature, the UNA has no specialists in extension, a field which contains a significant percentage of the total number of professionals in INTA and ICA.

The study of areas showed higher rates of turnover in the less traditional fields, such as economics, social sciences, and statistics, whereas the typically agricultural areas, such as soil science, agronomy, plant protection, extension, etc., occupy low or intermediate levels. This is reflected by the specific specialities, with their high rates of migration, particularly for agricultural economics, and to a lesser extent, in plant improvement and plant physiology.

The Causes of Migration

The causes of the migration were analyzed through an opinion poll conducted among the specialists who had resigned from the institutions from 1960 through 1978.

The central objective of the poll was to develop a hypothesis on the relative weight that the various causal factors exercise in the final decision to abandon an institution. In addition, the poll sought to generate information on occupational patterns and on the types of activities and organizations to which the specialists turned after leaving their original institutions.

It should be noted that while a high percentage of replies was received in Colombia, the validity of the

Table 5. For ICA, INTA, and UNA, areas as a percentage of total postgraduate personnel and rates of gross migration.[a,b]

	ICA		INTA		UNA	
	% of total	% emi-gration	% of total	% emi-gration	% of total	% emi-gration
Crop Production	*43.7*	*54.6*	*45.6*	*26.2*	*43.7*	*41.4*
Plant sciences	26.2	53.3	14.6	58.3	15.2	33.3
Plant protection	8.7	43.0	8.9	23.3	6.1	33.3
Soils	5.7	37.5	10.8	17.2	10.2	28.6
Agronomy	—	—	8.5	9.4	5.7	36.4
Agric. engineering	3.1	100.0	2.5	33.3	4.2	100.0
Forestry	—	—	0.3	0.0	2.3	200.0
Livestock	*23.3*	*37.8*	*20.4*	*14.2*	*11.7*	*55.5*
Agric. sciences	11.7	49.0	18.5	13.4	10.2	50.0
Veterinary sciences	11.6	26.4	1.9	20.0	1.5	100.0
Extension and Development	*14.6*	*35.5*	*12.1*	*52.0*	—	—
Other	*12.1*	*75.0*	*21.9*	*163.3*	*44.6*	*120.7*
Agricultural sciences	—	—	2.7	300.0	9.3	200.0
Economics and social sciences	9.0	100.0	15.7	226.0	20.5	100.0
Statistics	—	—	3.5	120.0	2.3	200.0
Engineering	—	—	—	—	9.1	60.0
Administration	3.1	28.6	—	—	3.4	800.0
Subtotal	93.7	48.8	100.9	44.7	100.0	70.3
Total	100	50.7	100	44.7	100	70.3

[a] Source: Developed from information provided by the institutions.

[b] Gross migration = (resignations)/(current staff).

information for Argentina and Peru is limited by the relatively low rates of response; only 23.4% and 17.7% of the total number of forms sent out were eventually returned. These rates of response hinder the possibility of interrelating the various causal factors in the shifting rates of migration during the period under study.

Table 6 summarizes 10 causal factors that the individuals identified as having exercised the greatest impact on their decision to migrate. The first important point that emerges from the survey findings is the low score assigned by surveyed INTA professionals to all the factors of resignation listed in the survey. This is consistent with the fact that, throughout the period, INTA experienced low rates of migration, with two isolated exceptions (1971 and 1976), which indicates the strictly transitory nature of the phenomenon. Although economic factors are mentioned as an important cause, they do not appear to be crucial, as they bear no clear correlation to annual rates of resignation. Other variables of a more institutional nature do show this correlation.

An example is "the presence of motivational mechanisms other than salary."[6]

The various causal factors showed comparable levels of relative importance for ICA in Colombia and the UNA in Peru. Again, complaints concerning salary and budget levels took a prominent position, receiving relatively high scores. Nevertheless, particular importance was ascribed to variables concerning the working environment, personnel policies (nonsalary mechanisms for motivation),

<hr>

[6] See: Ardila, J. et al. Sistemas Nacionales de Investigación Agropecuaria en América Latina: Análisis Comparativo de los Recursos Humanos en Países Seleccionados. El Caso de la Universidad Agraria de la Molina en el Perú. PROTAAL Document No. 49; and Ardila, J. et al. Sistemas Nacionales de Investigación Agropecuaria en América Latina: Análisis Comparativo de los Recursos Humanos en Países seleccionados. El Caso del Instituto Nacional de Tecnología Agropecuaria en Argentina (INTA). PROTAAL Document No. 48.

Table 6. Major causes of resignation for specialists with postgraduate studies from ICA, INTA, and UNA.[a]

ICA		INTA		UNA	
Cause	Score	Cause	Score	Cause	Score
A	2.37	A	1.40	A	2.73
E	2.27	C	1.40	E	2.53
C	1.90	F	1.33	C	2.40
D	1.78	I	1.27	I	2.26
B	1.68	G	1.13	B	2.00
F	1.53	J	1.07	D	2.00
G	1.45	B	1.07	K	1.86
H	1.43	K	1.00	H	1.80
I	1.43	H	1.00	L	1.60
J	1.41	D	0.93	M	1.60

[a] Source: Tabulated from a survey of former staff members who were with ICA, INTA, and UNA during 1960–78.

[b] A, salary; B, government support for the institution's functions; C, higher salary in another institution; D, internal working facilities; E, budgetary support for the institution; F, how much use the institution is making of its specialized professionals; G, managerial style; H, possibilities for professional advancement; I, presence of adequate nonsalary mechanisms for motivation; J, collateral government policies; K, institutional consistency in the objectives, functions, and activities by area of work for the postgraduate specialists; L, institutional acceptance of research findings; and M, ability of the institution to react to environmental changes.

and extrabudgetary support received by the institution from the government. As in INTA, the relationship between salary and resignation is weak, as there is no clear correlation between these two variables. This is not true for the more institutional variables, which consistently receive higher scores in years of high turnover.[7]

Thus, available evidence confirms the hypothesis that salary conditions are an important factor in the decision to abandon an institution, but they serve as a motivating force only when other institutional factors are present. In other words, the salary variable is important, but not in isolation. Rather, it serves only to complement and perhaps catalyze other elements that have inclined the decision-making process toward emigration. This becomes clear if we note that throughout the period, there have been high salary differentials between these institutions and the outside market; however, high levels of resignations are experienced only in certain

periods that do not necessarily coincide with the greatest differentials.[8]

Patterns of Occupation

Table 7 gives information on the types of organizations selected by outgoing personnel with postgraduate training. High percentages of these professionals leave research altogether. This phenomenon was most marked in INTA and ICA, probably due to the high concentration of research activities in these organizations. The situation was less clear at La Molina. Nevertheless, if the research and teaching complex is assumed to be the most natural sphere of activity for specialists with postgraduate degrees at UNA, a significant reduction could be seen.

These findings suggest that the migration process at the country level has produced net losses in the impact of the training programs on strengthening research capabilities. This is even more striking if it is noted that a large proportion of the resigning specialists move to international organizations and agencies: 11% for INTA, 35% for ICA, and 72.2% for UNA (Table 7).

Closing Comments

The resources available for this research project were limited. As a result, the final analysis suffered a number of shortcomings, and the findings and conclusions must be examined accordingly. In the first place, the study was restricted to a small group of countries selected on the basis of secondary information indicating that the problems of turnover among highly trained technical personnel had reached higher than normal levels. In the second place, it was impossible to examine the overall agricultural research system in each country, and therefore it was decided to concentrate the study on the most important or representative organizations. The selection included ICA in Colombia, INTA in Argentina, and the La Molina Agrarian University in Peru. Finally, only technical personnel with postgraduate degrees were considered, limiting the study to a single component of all the human resources involved in these institutions.

These limitations must be taken into consideration before making any generalizations. With this in mind, the findings of this study can be summarized as follows: (1) The changes in technical staff with postgraduate training tend to confirm the concerns that originally gave rise to the project. (2) Although

---

[7] Ardila, J. et al. (PROTAAL Doc. 48 and 49) and Ardila, J. et al. Sistemas Nacionales de Investigación Agropecuaria en América Latina: Análisis Comparativo de los Recursos Humanos en Países Seleccionados. El Caso des Instituto Colombiano Agropecuario (ICA) PROTAAL Document No. 47.

---

[8] Ardila, J. et al. (PROTAAL Doc. 48 and 49).

Table 7. Type of organization attracting specialists with postgraduate training upon resignation from ICA, INTA, and UNA.[a]

| | Percentage of specialists | | |
Type of organization	ICA	INTA	UNA
International organizations	21.6	8.3	62.2
National public institution	8.0	6.0	15.6
National private institution	29.7	30.5	—
International private institution	13.5	2.8	10.0
Teaching	13.5	25.4	12.2
Private business	10.8	17.1	—
Other	2.9	9.7	—

[a] Source: Tabulated from survey of former staff members who were with ICA, INTA, and UNA during 1960–78.

resignations have been high in Colombia and, to a lesser extent, in Peru, they are not the primary cause of the loss of personnel highly qualified for agricultural research. (3) Considerable effort has been made in the region to train human resources. These efforts were based primarily on available foreign funding because an adequate critical mass of personnel was viewed as a necessary condition for effective research. (4) The national postgraduate programs were founded in the late sixties, with the objective of guaranteeing that training efforts would not disappear after the large-scale loss of foreign funding. These programs have made important contributions to the process of training human resources, and in one case (Colombia), they have produced the majority of the personnel trained at the M.S. level. (5) In spite of the important contributions of the national programs and the central role they play in the long-term training of human resources, they have not obtained the national support they need. With the termination of foreign funding, they grew so weak as to nearly disappear. (6) The weakening of the national programs and the exhaustion of financial possibilities for pursuing training through scholarships and study abroad programs introduced a progressive deterioration of the overall staff. It should be stressed that staff reductions did not occur during periods of high turnover; rather, they were felt toward the end of the seventies, when the organizations had no easy alternatives for replacing lost personnel. (7) In the migration process itself, economic issues (personnel remuneration) are important, but not central, to the professional's decision to leave the institution. Much greater importance is ascribed to institutional variables involving the relationships between the specialist and the institution, as well as stability and placement in the working environment.

These findings point to institutional factors, including the nature and role of the research institutes, and the inability to consolidate national training programs, or to provide them with continuity, as central points in the search for alternative solutions to the problems under study.

The first factor, the nature and role of the research institutes, is perhaps the common denominator of the entire problem. It affects both sides of the staff loss problem: resignations and training to replace lost personnel. Nevertheless, because of its very nature, it is a topic that falls outside the realm of this document, as any relevant discussion would necessarily require an analysis of the degree to which the institutional research models have adapted to the characteristics of the agricultural sector in each country. Ultimately, it would lead to an examination of the role of research and technology in the complex of agrarian policies, a factor clearly external to the purposes of this study.

The second factor, training programs, is indeed a legitimate point of discussion, given the nature of this study. To provide a closing point for the discussion and a general framework for alternative solutions to the problems described above, it is important to reflect on the nature of the institutional context in which these programs unfold. The most important point is that the programs were not originally conceived as an integral part of overall training policies for the entire agricultural sector. Rather, the national programs were efforts by institutions that assumed training work as a function of their own needs and, in more than one case, as a direct result of outside initiatives.

This led to an isolation of the national programs, which became dependent on the needs and real potential of the institutions that generated them. This explains the breakdown that occurred during periods of institutional crisis, either structural (loss of budgetary support, modification of objectives and functions), as in the cases of ICA and the UNA, or in terms of policy shifts, as occurred with INTA.

Another important development, especially in the cases of ICA and INTA, is that in the absence of domestic training alternatives, the institutional programs, as well as the institutions themselves, became in fact the only source of trained personnel in the sector, for both public and private organizations. This informal function can be used to explain and justify certain high levels of turnover. Nevertheless, as the postgraduate programs received no formal recognition and reduced the internal profitability of the institutions, they cut into needed support for the institutions, especially at times of budget cutting, as occurred during the period under study.

In this context, a logical starting point for seeking solutions to the current situation would be a discussion of institutional possibilities for defining and implementing training policies on an overall level for the entire sector.

This paper is based on the findings of the project "National Agricultural Research Systems in Latin America: comparative analysis of human resources in selected countries," developed by IICA with support from the Rockefeller Foundation and the Ford Foundation. The authors would like to give full recognition to the following institutions that facilitated access to the basic information used in the analysis: the Colombian Agricultural Institute (ICA), the National Institute for Agricultural Technology (INTA), in Argentina, and the National Agrarian University of La Molina (UNA) in Peru. Particular mention should be made of the valuable cooperation of the members of the project's Advisory Committee, Armando Samper, Ubaldo Garcia, Jose Marull, Luis Marcano, and Luis Paz, who made considerable contributions to improving the analysis and interpretation of the information retrieved. We are also grateful for the participation of Norberto Reichart, Armodio Rincon, and Amador Merino Reyna in developing the case studies.

Development Strategy for Agricultural Research Manpower in Indonesia

Sjarifuddin Baharsjah[1]

Compared with other countries, Indonesia has less university-trained scientists. In 1975 there were only 80 scientists in all fields for every 1 million Indonesians. The number of agricultural scientists was even lower because higher education in agriculture only began to produce graduates in significant numbers during the last 15 years, although university level training in agriculture in Indonesia was formally decreed in 1941. At that time, faculties of engineering, medicine, and law were already established in the country. On the other hand, agricultural research started quite early. The Botanic Garden in Bogor was established more than 100 years ago. Estate crops research was established to cater to the needs of the plantations. Later, when plantations in the outer islands were badly in need of labour, the colonial government supported them by launching a transmigration/resettlement program. As new land had to be opened for these resettlement projects, a soil research institute was needed and later established. Research institutes for rice and other food crops were also established. Practically all researchers in these early institutes were expatriates, mostly Dutch. Therefore, an acute lack of agricultural scientists and researchers was felt when an abrupt exodus of these expatriates occurred in the late fifties.

In 1967, agricultural development started in earnest and very soon the need for new, and superior, technologies was felt. Unfortunately, research had not been able to attract graduates and scientists to replace the Dutch researchers and the need for this new technology could not be met adequately.

By 1975, when the Agency for Agricultural Research and Development (AARD) was established,

it inherited 13 research institutes with 744 research workers of whom 17 held a Ph.D./Dr, 44 a M.S., 470 were university graduates, and the remaining 206 were technicians. A measure of the failure of research to provide incentives to attract capable graduates was the fact that between 1968 and 1976 the number of researchers increased exactly at the same rate (11%) as the increase of university graduates for the period.

The number of research personnel in 1976 in different age groups at the Bogor Estate Crops Research Institutes clearly shows the gravity of the problem: over 50 years old, 4; 45–50 years, 13; 40–45 years, 20; 35–40 years, 15; 30–35 years, 3; and under 30 years, 1.

It seems that planned recruitment in the institute ceased about 20 years ago, which was approximately the time of the exodus of the expatriate experts. Addition of new personnel after that time did not show any consciously planned replacement or expansion program. Other institutes, notably the Center Research Institute for Agriculture (CRIA), were in a somewhat better position. This was mostly due to the exceptionally forward looking leadership of CRIA and to cooperation between CRIA and the International Rice Research Institute (IRRI) in the Philippines.

Agricultural Research Manpower Development Program

Two events provided the opportunity for planned agricultural research manpower development in Indonesia: (1) the establishment of AARD in 1975; and (2) the establishment of a graduate school at the Bogor Agricultural University (IPB) in 1974.

Prior to AARD, the research institutes, which were organized along commodities or groups of commodities, were under the various Directorate Generals that were responsible for the development of the particular subsector. Thus, CRIA belonged to

[1] Senior Lecturer, Department of Rural Sociology and Agricultural Economics, Bogor Agricultural University (IPB), and Director, Center for Agro-Economics Research, Agency for Agricultural Research and Development (AARD).

the D.G. Food Crops Agriculture, The Estate Crops Research Institutes belonged to the D.G. Estates, etc. Matters concerning research personnel were part of the D.G. personnel policy, which was not necessarily supportive of efforts to increase research capacity. This changed when AARD was established. As a national agricultural research program began to develop, the need for a research staff development program became apparent. AARD soon adopted a personnel policy that would support the program.

The graduate school at IPB was opened 1 year earlier than planned. This was due to the large and urgent demand in agriculture for scientists and researchers with advanced degree level training. In 1972, IPB began to restructure its curriculum from a single 6-year engineer's training course to a three tier program: a 4-year first degree or S(I) level, a 2-year S(II) or M.S. level, and a 2–3 year S(III) or Doctoral level. The reason for the change was that the old 6-year engineer program was quite unproductive, in part, because it attempted to train every student to become a scientist or researcher. The S(I) provides students with the opportunity to graduate and enter the job market as skilled practitioners. Only those who qualify continue into the S(II) or M.S. level to become scientists and researchers.

Two kinds of opportunities were provided by the opening of the IPB graduate school. First, this was the first time that rigorous training at the M.S., and later the Doctorate level, in the various fields of specialization in agriculture became available within the country. It is now possible to send three trainees to the IPB graduate school at the same cost as one sent abroad. Furthermore, it became easier to link the trainees' M.S. thesis and Doctorate dissertation research with the sending institutes research program, making the research more relevant. Second, the restructuring of the IPB program enabled the institutes to recruit and appoint promising S(I) graduates and later send them, in their capacities as the institutes own staff, back to school at IPB to be trained in research at the S(II) and S(III) levels. This solved a number of personnel status problems and enabled AARD to provide fellowships.

The Strategy

As was stated earlier, even before AARD, the CRIA leadership was pursuing a research staff development plan and was quite active in recruiting. Unfortunately, other research institutes were not as progressive. CRIA's success was to a significant extent due to its ability to cut through established personnel policies such as the adherence to seniority for training opportunities. There was a need to establish new policies at the AARD level that were conducive to, and supportive of, recruitment and training of potential researchers in all research institutes. Centralized planning was also necessary to relate research manpower development, and the allocation of funds supporting it, to the national agricultural research program. Priorities in the research program had to be clearly reflected in the manpower development program.

Targets

The 1968–76 annual rate of increase of 11% in researchers was not adequate. It was decided that between 1976 and 1985, 2000 graduates from agricultural universities and faculties needed to be recruited and trained at the S(II) (M.S.) and S(III) (Doctorate) levels. Because about 400 other trainees were already in the institutes, the target implied an annual rate of increase in researchers of about 15%. This target was set after considering the supply of new graduates. The number of graduates expected in all fields of studies by 1985 was officially estimated at about 200 000 persons, of which about 30 000 (15%) will be in agriculture.[2] Although the 2000 new recruits that are targeted will be far below the 11% (of the 30 000 agricultural graduates) who are expected to be attracted to join research, it was recommended that the research institutes pursue a vigorous recruitment program. There were two reasons for this: as development is successful, new and better paying job opportunities will be created and fewer people may be attracted to research; and it is important to recruit only those who are most capable among the graduates.

Priorities

Because of the general shortage of researchers in all fields, in every research institute priority was given in the first years of the program to those candidates with the highest academic standing. Starting this year, after more than 300 have been recruited and sent for graduate training, a new set of priorities will be used that is formulated to support the national research program. As new laboratories and research facilities are built in the regions, it is necessary to attract the required researchers. Thus, willingness to work in these new research complexes, some in relatively remote places, is given top priority. High scores are also given to those studying in unpopular fields. In every case, however, academic performance is considered.

---

[2] Makagiansar, M. Memorandum to Akhir Jabatan, Jakarta 1976.

In-Country Versus Foreign Training

As long as a field of study is available in a graduate school in Indonesia, in-country training is preferred. It has been possible to provide the research institutes with funds earmarked for support of M.S. thesis and Doctoral dissertation research conducted by their trainees. Wherever possible, a qualified senior researcher from the institute becomes a member of a trainee's academic advisory committee.

Provision of AARD Fellowship

With a comprehensive 10-year research development program, it was possible for AARD to secure the funds from its development budget and from IBRD loans. With these funds, AARD provides fellowships that include tuition, stipend, book and travel allowances, and research support. Because trainees are fully supported, the research institutes are requested to relieve them of other assignments.

Utilization

To date, of a total of 304 trainees sent for graduate level training in the country and abroad, 72 have completed M.S. and Ph.D./Dr degrees. They have returned to their respective research institutes. When they were being trained, steps were taken to ensure they would be utilized when they returned. These steps included maintenance of communication between the trainee and his research institute. In addition, AARD tries to remove budget constraints so that trainees do not complain that they are unable to conduct research due to unavailability of funds. A more difficult problem is the provision of adequate support staff. There is a general shortage of qualified and well-trained technicians in the research institutes. In recent years, the private sector has shown an increasingly strong demand for laboratory and field technicians, programers, and well-trained secretarial personnel. This demand is backed up by the private firms' willingness to pay good salaries. The research institutes, which are bound by public regulations, cannot compete with the private sector in this respect. Utilization is also assisted by the construction of new laboratories and other research facilities.

Promotions

Even before AARD, researchers enjoyed a system of dual promotional arrangements. As government officials they are promoted under the general regulations in which position and seniority are emphasized. As scientists they are also evaluated for their accomplishments and much less emphasis is given to seniority. Functional promotions are recognized beyond AARD, and also by the Indonesia Foundation of Sciences (LIPI) and the universities. They also carry salary increases.

Problems

Although some measure of success has been achieved, some problems remain. Several can be solved by AARD and its research institutes, others need to be tackled in cooperation with other institutions, such as the universities.

(1) Allocating researchers in the institutes versus sending them to school. As appreciation for research rises so does the demand for more research. The research institutes face an allocation problem, i.e., whether to retain their capable young researchers or send them to school. Often the tendency is to nominate their older but less capable staff for training. AARD seeks to alleviate this problem by: (a) developing more understanding and cooperation with the institutes; (b) setting a general maximum age limit for those eligible for training with AARD support; and (c) insisting that a minimum academic requirement be met by all candidates.

(2) Recruiting for unpopular fields of study and for assignment in the more remote places. A number of important fields of study remain unpopular with the candidates. These include the basic sciences, genetics, agrometeorology, aquaculture, marine biology, and sociology. Because trainees are nominated by the research institutes, AARD attempts to solve this problem with the directors of the institutes by means of talent scouting with the collaboration of the universities. AARD also assigns high priorities to those willing to study in these unpopular fields and who are willing to be assigned in the new facilities situated in the more remote places. An additional incentive is training abroad because in many cases no graduate training in these fields is available in Indonesia.

(3) Selection of universities. Currently only three universities in Indonesia are acredited to offer graduate level courses in agriculture. Because the demand for these programs is very large, these universities are conducting their program at full capacity. IPB, for example, has a graduate enrollment of over 600 in both M.S. and Doctorate levels and faces difficulties in enlarging any further. No doubt, the AARD staff development program has created momentum for other universities to begin offering graduate programs. AARD will maintain its policy of training only at accredited universities because it feels this is the best way to help develop strong graduate programs in Indonesian universities.

Manpower Developments for Agricultural Research in Bangladesh

S.M. Elias[1]

Agricultural production in Bangladesh continues to lag behind requirements. The second five year plan (SFYP) for the period 1979–80 to 1984–85 rightly seeks to achieve rapid growth in agricultural production. In view of the importance of agriculture, which provides 55% of the gross national product, more than 85% of total employment, and earns over 80% of foreign exchange, the desire to have rapid growth is just in time. To increase the production as well as income of rural people, a comprehensive agricultural research plan is essential. The Bangladesh Agricultural Research Council has prepared a national agricultural research plan that defines the program priorities for agricultural research during the second five year plan. In this plan, both commodity and noncommodity programs have been given priority. Research institutes that are actively engaged in agricultural research have devised their own programs for achieving the objectives of the five year plan. However, to conduct such a program, more scientific personnel, both in number and in quality, are required. The objectives of this paper are to analyze the requirements for scientific manpower and to assess the need for training to carry out the program for the plan period.

Manpower Requirements

The institutes directly engaged in agricultural research work, which includes research on crops, livestock, forestry, and fisheries, are: Bangladesh Agricultural Research Institute (BARI); Bangladesh Rice Research Institute (BRRI); Bangladesh Jute Research Institute (BJRI); Forest Research Institute (FRI); Institute of Nuclear Agriculture (INA); Tea Research Institute (TRI); Sugarcane Research Institute (SRI); Fisheries Research Institute (FIRI); Livestock Research Centre (LRC); and Bangladesh Agricultural University (BAU). Over 1500 scientific personnel are available in these institutes. Table 1 shows the existing manpower classified by academic qualification and research institute.

In view of the challenging job to be done in the second five year plan, each of these institutes has assessed its additional requirements for manpower (Table 1). A total of 933 scientists will be required during this plan period to fulfill the objectives of the plan. Of this number, over 350 scientists are required to be Ph.D. degree holders and over 460 scientists are to be Masters degree holders.

Source of Manpower Resources

It is recognized that the Bangladesh Agricultural University and the College of Agriculture are the source of most trained manpower in the country. Table 2 shows the number of B.Sc. students graduated during 1975 and 1976 in the different faculties. It also shows the number of M.Sc. students graduated up to 1976 and the number enrolled in 1977 and 1978. In 1975 and in 1976, 472 and 529 students received a B.Sc. and 153 and 193 students were enrolled as M.Sc. students in 1977 and 1978, respectively. This university also offers Ph.D. degrees. From 1972 to 1980, three students earned a Ph.D., and at present five students are enrolled in Ph.D. programs in four different disciplines. Agricultural extension and supply services demand several times more agricultural graduates than research. In this context, it can be stated that such a large requirement for scientific manpower (i.e., 933) during the plan period cannot be supplied by one agricultural university. It is essential to train and develop existing, as well as new research personnel, if the plan's objectives are to be fulfilled.

---

[1] Head, Division of Agricultural Economics, Bangladesh Agricultural Research Institute, Joydebpur, Dacca, Bangladesh.

"""

Table 1. Existing scientific manpower and the projected requirements for the second five year plan period.[a]

	Ph.D.		M.Sc. (Ag)		B.Sc. (Ag)		Total	
Institute	Existing	Req'd	Existing	Req'd	Existing	Req'd	Existing	Req'd
BARI	15	90	525	55	60	55	600	200
BRRI	23	18	105	84	3	8	131	110
BJRI	8	15	125	50	28	12	161	77
FRI	6	15	68	30	24	1	98	46
INA	7	5	42	12	18	—	67	17
TRI	1	17	20	36	9	3	30	56
SRI	4	11	12	33	4	—	20	44
FishRI	—	10	15	15	21	12	36	37
LRC	2	15	21	35	74	10	97	60
BAU	78	158	185	118	11	10	353[b]	286
Total	144	354	1118	468	252	111	1593	933

[a] Source: BARC, National Agricultural Research Plan, December 1979.
[b] After adjustment.

Table 2. Number of B.Sc. graduates in 1975 and 1976 along with number of M.Sc. graduates up to 1976 and number of M.Sc. students enrolled in Bangladesh Agricultural University in 1977 and 1978.[a]

	B.Sc.		M.Sc.		
			Graduated	Enrolled	
Discipline	1975	1976	up to 1976	1977	1978
Agriculture Faculty	226	241	636	47	95
Animal Husbandry Faculty	74	74	96	8	31
Veterinary Science Faculty	52	71	104	19	20
Agricultural Economics Faculty	29	53	190	52	26
Fisheries Faculty	29	34	55	34	12
Agri. Engineering Faculty	62	56	5	1	—
Total	472	529	1086	161	184

[a] Source: Bulletin from Bangladesh Agricultural University, 1979.

Need for Manpower Development

Trained scientific manpower is the foundation on which the superstructure of research is built. The quantity of research depends primarily on the skill, knowledge, and competence of researchers. Quality of research depends mainly on what training the scientists have acquired in their own discipline.

New researchers, fresh from university, need to be oriented so they know the objectives of the institute where they work and know the specific job they are to perform. Junior researchers must know the past as well as the recent development of their research subject; senior researchers must observe recent developments in different countries and accordingly apply the most appropriate techniques to their own country; and research managers must know modern management techniques for efficient research management at their institute. Thus these scientists need to be trained in their own fields of specialization. At present, there is already a shortage of qualified scientific personnel to fill the existing research program. If the research objectives of this five year plan are to be achieved, a comprehensive manpower development program needs to be developed for all research institutes to increase the quantity and uphold the quality of the scientific personnel.

Three kinds of training have been identified for agricultural research personnel: (1) on-the-job training; (2) academic training; and (3) short-term training. All three kinds of training are equally important for the development of the scientific manpower of a research institute. Academic training helps to form the sound base for a researcher; short-term training helps to develop skill, knowledge, and efficiency. It is necessary that trained personnel be provided with facilities to demonstrate their compe-

Table 3. Relationship between manpower allocation, value of production, and area covered for different commodities.[a]

Commodity program	Manpower requirements		Value of production		Area covered (1978–79)	
	1981–85 (Man-years)	% of total	Million taka	% of total	Thousands of acres	% of total
Cereal	192	23.8	689	9.5	850	21.3
Pulses	86	10.6	409	5.7	842	21.1
Oilseeds	148	18.3	246	3.4	712	17.8
Fibres/Tobacco	87	10.8	348	4.8	140	3.5
Fruits and palms	76	9.4	1932	26.8	352	8.8
Vegetables	72	8.9	1599	22.2	296	7.4
Spices	59	7.4	887	12.3	385	9.6
Roots and tubers	87	10.8	1101	15.3	419	10.5
Total	807	100	7211	100	3996	100

[a] Sources: BARI, Second Five Year Research Programme, 1980; Bangladesh Bureau of Statistics, 1980.

tence and training. This requires a long-term manpower development program for each discipline and for each level.

Manpower Allocation in Different Programs

A serious imbalance is observed among different sectors and disciplines of agriculture. Eighty-one percent of the total value of all agricultural production is supplied by the crop sector, but only 78% of the total agricultural research manpower is engaged in this sector. Similarly, the value of production contributed by livestock, fisheries, and forestry is 7, 9, and 3%, respectively, while about 12, 4, and 6% of agricultural research personnel are engaged in these sectors.

Thus, the crop sector did not get its due share of research manpower in relation to its contribution to the total value of agricultural production. Similarly, a close look into the crop sector reveals that an imbalance of allocation exists in the different institutes with respect to manpower.

Bangladesh Agricultural Research Institute (BARI)

BARI, because it is the largest research institute in the country and deals with a large number of crops,[2] will be given special attention to observe its

---

[2] Except for rice, jute, sugarcane, and tea, for which there are separate institutes, BARI deals with research on all other crops.

proposed allocation of manpower in different programs to achieve the objectives of the second five year plan.

BARI has developed 17 multidisciplinary research programs. Some of these programs are commodity oriented while others are noncommodity programs. Manpower requirements for the five year plan period for these programs and the estimated value of production of each commodity are shown in Table 3. Although the total value of production of crops like fruits, vegetables, spices, potato, and root crops is more than the value of field crops like cereals, pulses, or oilseeds, manpower requirements for those horticultural crops have been planned to be lower than that of field crops. In other words, value per scientist appears to be much higher in the case of horticultural crops. However, if area covered is considered instead of value of production, the picture becomes altogether different. In that case, pulse crops need much more scientific manpower than is planned: 21.1% of the reported crop area is devoted to pulse crops but only 10.6% of the total required manpower has been assigned to the pulse research program. But, area under a crop also cannot be considered as the single criterion because there are many other constraints. Food habits of the rural mass, principal food crop of the majority of the people, nutrition provided to the common people, and foreign exchange earnings are just some of the criteria that must not be neglected when scientific manpower are to be allocated to different commodities.